Edwin J. Clapp

The Port of Hamburg

Edwin J. Clapp

The Port of Hamburg

ISBN/EAN: 9783955640569

Auflage: 1

Erscheinungsjahr: 2013

Erscheinungsort: Bremen, Deutschland

@ EHV-History in Access Verlag GmbH, Fahrenheitstr. 1, 28359 Bremen. Alle Rechte beim Verlag und bei den jeweiligen Lizenzgebern.

Row of buildings of the Free Port Warehousing Company.

THE PORT OF HAMBURG

BY

EDWIN J. CLAPP

DEPARTMENT OF TRADE AND TRANSPORTATION IN NEW YORK UNIVERSITY

Author of
"THE NAVIGABLE RHINE."

NEW HAVEN: YALE UNIVERSITY PRESS
LONDON: HENRY FROWDE
OXFORD UNIVERSITY PRESS
MCMXII

TO MY FRIEND
WILLIAM H. TAYLOR
OF NEW YORK

PREFACE

THIS book was written with the conviction that the much-needed modernization of our ocean and Great Lakes terminals must be along the lines followed in Hamburg, and that river transportation in America, if it is ever to be resuscitated, must be modeled on that of the great German streams, the Elbe and the Rhine. Yet no attempt is made to prescribe what shall be done in America; this is an investigation, not a program nor a prophecy. It has aspired to be the sort of study which must precede any sane program.

Books on ocean and inland waterway transportation are neither many nor good. The bibliography here given, if examined, would satisfy nobody. The field is one where personal investigation will probably long continue to be more productive than the reading of many books. It is hoped that the illustrations will help the reader understand what is new and strange to him, better than mere description could.

Because of the unsatisfactoriness of the literature, Wiedenfeld's great book, "Die nordwesteuropäischen Welthäfen," stands out with particular distinctness. Like all students of European seaports, I must express my acknowledgments to him, as well as to Cords for his elaboration of Wiedenfeld's method of handling the statistics of German inland trade. The Hamburg-American Line has shown me courtesies; it loaned many of the photographs here used. The success of a personal investigation depends on the coöperation of many individuals, too many for me to thank in detail. Above all else I am indebted to the active aid and constant encouragement of my wife.

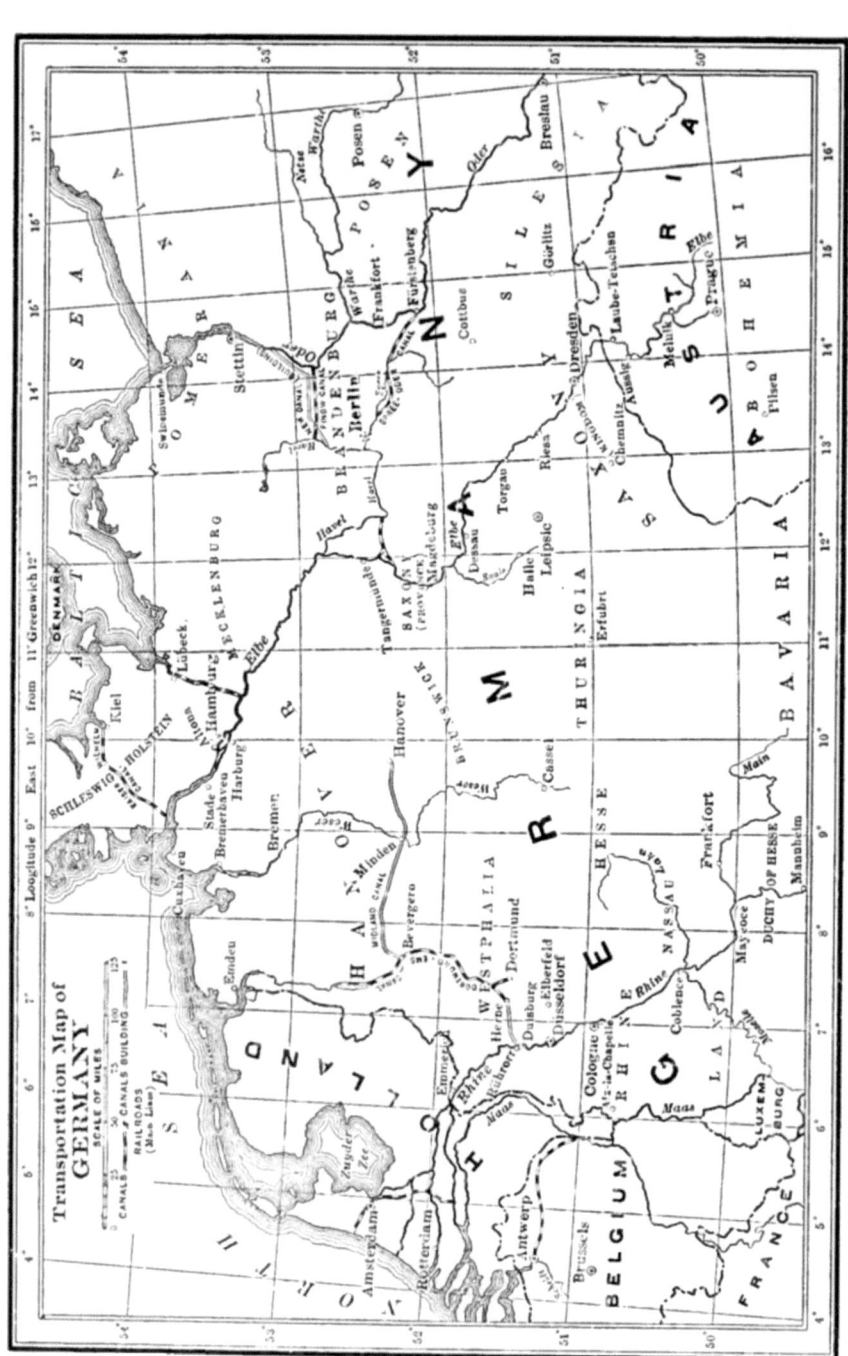

CONTENTS

PAGES

INTRODUCTION. Nature of a Great Seaport . . 1-14
Great seaports. Dependence on a hinterland. Equipment of a great port. Channel to the sea. Harbor facilities. Contact between rail and ocean carriers. Facilities for transfer of freight between ship and inland barges. Railway lines to interior, inland waterways, ocean steamship lines. The utilization of this equipment. Harbor dues. Railway tariffs. Utilizing the waterway. Activity of merchants. Active extension of steamship lines. Place of seaport in national life. Hamburg's equipment and her use of it. Similarity between her conditions and ours. Open basin type of harbor. Freight-handling machinery. Contact between rail and water carrier. Utilizing the Elbe. Suitability of a study of the port of Hamburg.

CHAPTER I. Development of Hamburg's Hinterland 15-28
Nature of Hansa ports and their trade. Fall of the Hansa. Remains of its trade. Hamburg's new position. Beginnings of free trade on the sea. Unifying Germany. Formation of the Empire. Economic development of Germany. The new industrial state. Imports and exports. Industry supplants agriculture. Proportion of foreign trade carried on by sea. Hamburg's part in German foreign trade. Hamburg's dependence today on her hinterland.

CHAPTER II. The Channel to the Sea . . . 29-42
Draught of Hanseatic vessels. Of modern liners. Bulk goods in commerce. Importance of deep channel for steamers. Advantages of inland seaports. Types of modern steamers. Hamburg's aim. Nature of channel improvement. The present channel. Future work. Cost of channel. Mediæval tolls. Present channel dues. Principles of financial policy. Hamburg's advantage over its rivals. Summary.

CHAPTER III. Port Facilities 43-76
The harbor before 1850. Ancient methods of discharging and warehousing. Steamers and their new demands. Con-

CONTENTS

PAGES

struction before 1882. Hamburg enters the Customs Union. The new Free Port. Its commercial advantages. Its manufacturing advantages. Its maritime advantages. The belt canal. New piers of Hamburg-American Line. Pier sheds. Unloading liners. Museum model in Berlin. Use of midstream mooring posts. State piers and leased piers. Handling coal. Area of the harbor. Draying in Hamburg. Lightering. Handling barges. The railroad terminal. German forwarders. Free Port Warehousing Company and the warehousing business. Emigrant village of the Hamburg-American Line. Harbor works in Cuxhaven. Its abandonment except as port of call. Cost of harbor construction. Fiscal policy. Hamburg compared with other ports. Speed of discharge is the essential. Summary.

CHAPTER IV. Hamburg's Oversea Lines . . 77-109

The importance of direct regular German lines. Emancipation from English carrier and middleman. The new steamship line and the old merchant carrier. Identity of interest between Hamburg's shipowners and merchants. Beginnings of the Hapag. Former predominance of the steerage business. Its extent today; its significance. Passing of the sailer. Hapag's profits, 1860-70. Boom after 1871. Hamburg South American, Kosmos and Kingsin lines. Rate wars, 1872-76. Development after 1886. Hapag's agreement with Kosmos Line, absorption of Kingsin Line. Woermann Line. Temporary Hapag Line New York-Levant. Hamburg-Australian Steamship Company. German East Africa Line. Union of Hapag and Hamburg South American lines. Further development of Hapag. Its service. New York-West Indies. Port Stilwell-Orient, Argentine-Genoa, etc. Hamburg's short traders. Size of different Hamburg lines. The world's largest steamship lines. Hapag and the American merchant marine in the foreign trade. Advantages in diversity of Hapag's services. Examples. Profits from wars. Present surplus of tramp steamers. Summary.

CHAPTER V. Hamburg's Shipbuilders and Merchants. State Aid to the Merchant Marine 110-139

Need of repair facilities in ocean ports. Hamburg's facilities. New center of German shipbuilding. German navy

CONTENTS

PAGES

is home-built. The Lloyd's subsidy in 1885. The Lloyd and German yards. The Hapag and German yards. Use of German steel now. Ship types in the merchant marine. Type preferred by the Lloyd; by the Hapag. The Hapag's bid for passenger traffic. Its third class. Hamburg's merchants, where they came from. Retention of ancient virtues. The seaport merchant and exports, bulk imports, valuable imports. Hamburg as a market for colonial products. London's advantage. Hamburg's field abroad. German government and German exports. Forms of ship subsidy. Mail contracts. Trade subsidies. German mail pay. The Lloyd's subsidy; its conditions and its success. The East Africa subsidy; its success. Subsidies need the right merchants at home. Subsidies in foreign marines. German preferential railway rates. Summary.

CHAPTER VI. Waterways and Railways. . . 140-172

Seaport and waterway. Hamburg and its inland waterways. Competition with Stettin for East Germany. Preponderance in Central Germany. River improvements. Modernization of river craft. Types. Organization of transportation companies. United Elbe Navigation Company. Barge terminal facilities in Hamburg; in river ports. Rate policy of Prussian railways. Saxon and Austrian railways. Water freight rates. Combined rail-and-water rates. Hamburg and German export and import rates. Their effectiveness in South and West Germany. German Levant and East Africa tariffs. Their aid to German exporters. Import rates. Summary.

CHAPTER VII. Shipping and Commerce in Hamburg. 1907 173-192

Hamburg the first port of Europe. Shipping in German ports, by flags. North Sea and Baltic ports. Hamburg's predominance. Hamburg's shipping since 1850. Dominance of German flag. Hamburg's merchant fleet and line connections. Analysis of Hamburg's shipping. Commerce in 1907, growth. Nature and increase of imports, of exports. Exports and imports by leading countries. Re-exportation. Importation and the hinterland. Summary.

CONTENTS

PAGES

CHAPTER VIII. Hamburg's Commerce with its Hinterland. 1907 . . . 193-210

Statistics at our disposal. Hamburg's total commerce with the interior. Analysis by carriers and goods. Proportion of Elbe and railway in shipments to, and receipts from, East and Central Germany. Victory of the waterways. Competition with Bremen in East and Central Germany. Competition of foreign ports in West and South Germany. Summary.

LIST OF ILLUSTRATIONS

PAGES

Row of buildings of the Free Port Warehousing Company	Frontispiece
Transportation map of Germany	ix
Hamburg freighter, being served by barges and lighters	13
The Steckelhorn, one of Hamburg's ancient canals	26
The "Kaiserin Auguste Victoria"	39
Hamburg freighter at a pier, discharging	51
Pier cranes discharging cargo at Hamburg	63
Church in the emigrant village at Hamburg	75
The "Deutschland"	89
View looking into the main basin of the Hamburg-American Line	104
Launching of the "Fürst Bülow"	117
The "Imperator," the Hamburg-American's new monster liner	130
River terminal at Torgau on the Elbe	143
River scene on the Elbe at Magdeburg	143
River scene on the Elbe at Laube-Tetschen	156
Pier crane in the river port at Torgau on the Elbe	156
The levee at Saint Louis	169
A cotton landing on a tributary of the Mississippi	169
Loading a trainload of cable direct from cars into steamer	182
Elbe barges being discharged at a steamship pier at Kuhwärder	195
Map of the harbor at Hamburg	214

THE PORT OF HAMBURG

INTRODUCTION

THE NATURE OF A GREAT SEAPORT

A GREAT seaport is a country's right hand extended to foreign lands, offering them our products and requesting theirs. It is the focus of a variety of lines of communication: ocean steamship lines engaged in the coasting and the foreign trade, inland waterways and railways. Its function is to bring these lines into contact and to enable them, with the least possible friction and loss of energy, to effect the exchange of their burdens. In a seaport are knit together the bonds that unite the nations in a network of ever increasing complexity; the seaport is the highest expression of that new phenomenon of the nineteenth century: world-wide trade. It is the great clearing house for the material goods of international commerce. It is the heart of a country's commercial life, drawing off the sluggish flow of surplus inland production and sending back through the arteries of traffic the life-giving currents of foreign trade.

The first requisite for the existence of a great seaport, nowadays, is the existence of a hinterland interested in foreign trade. London is a last, waning example of the old order of seaport which throve primarily because it was the entrepôt of nations and whose business consisted mainly in transshipment and re-exportation. Today these nations have their own trade connections with lands across the sea and need no entrepôt to mediate for them. A modern port does not import for foreign countries nor does it import for itself, just as the heart does not draw

THE PORT OF HAMBURG

through the veins blood for its own particular use. It imports for a hinterland. Similarly, it sends abroad the sum of the exports of inland points. The finest harbor in the world cannot become a great seaport if it is located on the coast of Norway or the island of Madagascar. London and Liverpool developed so much earlier than all other ports because England long remained the entrepôt and workshop of the civilized world. The development of the port of Hamburg has been parallel with the growth of Germany's foreign trade. Hamburg has caught up with London and Liverpool now largely because Germany is crowding England on the markets of the world. If once this basic condition is present—a hinterland eager for foreign trade—the port's greatness depends on its equipment and the use it makes of that equipment.

By the equipment of a port, in the larger sense, we mean a number of things. It is first of all necessary to have a channel from the sea deep enough to let the largest vessels come up, if possible, even at low tide. Speed is life to the ocean liner and several hours every trip spent waiting at the harbor bar may mean the loss of half a trip a year and may result in loss instead of profit for the year's work. To provide a channel for the mammoth vessels of today, expensive dredging is necessary. What a problem this channel dredging is becoming is apparent when we consider that most of the older seaports were located at that point where sea and river navigation met. The small mediæval ships could penetrate far inland; for instance, Hamburg lies eighty-five miles up the Elbe.

Once the ship has entered from the sea, she needs a harbor space large enough for her to go to her berth without disturbing other vessels that are moving, at

INTRODUCTION

anchorage or at their piers. There must be enough piers to accommodate the regular liners and such free lances as may be expected to arrive. On the piers should be substantial sheds to shelter inbound and outbound cargo and there should be freight-handling machinery on the pier's edge to expedite the loading and unloading of vessels. Warehouses should be at no great distance from the piers, especially the warehouses that shelter the goods likely to be re-exported. There must be shipyards and docks to rebuild, repair and overhaul ships that are in need.

Railroad tracks must run out on the pier so that goods can be sent inland by rail directly, without the expensive operation of draying them across the city to the freight-receiving station of the railroad. If there is more than one railroad serving the port, there should be a harbor belt line, owned by the city or by the railroads jointly, which would classify and deliver cars between any pier and any road. The art of transportation has nowhere progressed so slowly as in this matter of terminals: in the provision for a cheap, expeditious, frictionless transfer of goods from one vehicle of transportation to another. London is an exceptional case in that its huge population of seven and one half millions demands so much from abroad that the greater part of London's imports are destined for this population or for re-exportation. Elsewhere, most of the freight arriving at a seaport goes inland, usually by rail, and therefore has to bear the burden of terminal charges that are often unnecessarily high. Under the competition now raging for every rich hinterland, the prime need of an aspiring competitor is to draw to itself combined rail and ocean transports by

THE PORT OF HAMBURG

cheapening the transfer between boat and car. It is partly the intensity of competition for a rich territory that has forced Hamburg up to her present high efficiency. In the space of 300 miles, from Antwerp to Hamburg, there are five great seaports: Antwerp, Rotterdam, Amsterdam, Bremen and Hamburg, all competing for the oversea trade of industrial central Europe. One might expect them to be the best built and managed ports in the world, and so they are.

Besides harbor lightering, which provides for the collection and distribution of small shipments between ocean and inland carriers and the transshipment of cargo from one vessel to another—for instance, from ocean-going liner to coasting vessel—there should be provided contact between the seaship and whatever inland craft ply on the river above our port. If the seaship brings cargo which is to be discharged solely into up-river barges, the ship need not take up room at a pier; she can tie up to mooring posts in midstream or wherever else she will be out of the way, and there discharge into the barges. Barges having to do with ships lying at piers should be compelled to exchange all small shipments with them by means of lighters; it is impracticable to tow into the slip every big barge that has twenty to thirty tons of cargo to get from the liner. Capacious basins must be provided in the port for the river barges and tugs where they can lie undisturbed and undisturbing while waiting for their ship to come in or while waiting for an upstream tow to form.

All this is recognized as a port's equipment. But in the larger sense this equipment is far more. It includes a great network of railways stretching across a continent and collecting treasure to enrich the port. Its equip-

INTRODUCTION

ment includes the river on which it lies, with all its tributaries and connecting canals, affording for bulk goods such cheap transportation as the railroads cannot give. Lastly, the equipment of a great seaport includes a vast system of lines on the ocean: transoceanic steamship lines, steamship and barge lines that ply coastwise and to nearby foreign countries. The more numerous these lines of communication are, the greater is the transfer of goods between them, the greater is the volume of trade that flows through the seaport.

Yet a great seaport is not a mere passive tool, a sort of international freight yard where goods are switched about. It is more like a living organism, with a mind and will of its own to use in the struggle for existence. As important as the technical equipment of a port is the use that the port makes, or is allowed to make, of the equipment. Even if the channel and the harbor are perfect, their usefulness may be curtailed by the attempt to collect channel and harbor dues high enough to cover not only the cost of maintenance but also interest on all money expended on them. That is, these expenditures may be considered as an investment which must be made to pay rather than as public improvements designed to promote the free flow of commerce and create wealth that can be better taxed at another point. The latter point of view is the farsighted one. If the port chooses, it can even go so far as to remit all channel and harbor dues for ships engaged in commerce with countries where its merchants or steamship lines are laboring to get a foothold.

The services rendered the port by the railroads that enter it are not measured by the length of their tracks

THE PORT OF HAMBURG

but rather by the policy they pursue: the rates they give. Low rates for long hauls can extend the hinterland of the port; special rates for raw materials imported and manufactures exported can build up in that hinterland industries and an industrial population dependent on constant connection with oversea. Railroads and their allied steamship lines can prorate and give especially low tariffs for export to foreign markets where international competition is intense.

Perhaps more important than in the preceding cases is the use to which internal waterways are put, as distinct from the mere existence of those waterways. There can never be a heavy traffic on a waterway that carries merely the goods to be exchanged between a seaport and points on the waterway. A heavy traffic can arise only when these river points collect and distribute the foreign trade of a large territory. The Mississippi can never have more than a desultory river trade if it carries only the goods exchanged between New Orleans and Vicksburg, St. Louis, Memphis and St. Paul. It can have a very large river trade if Memphis can collect and distribute oversea shipments for all Tennessee, St. Louis for all Missouri, St. Paul for all Minnesota. That would give ample cargo for barge trains for the seaboard. Hamburg and New York would never be first-class seaports if dependent on the exchange of goods with each other; they collect and distribute each for a wide hinterland. Nothing but a similar activity can make a great river port or a great river traffic. This is a fundamental truth in waterway transportation that is too often neglected.

If waterways are to render this service of providing

INTRODUCTION

cheap transportation between a large territory and the seaboard, there must obviously be coöperation between them and the railways, for waterways do not, at best, reach more than a small number of the destinations of imported goods or of the sources of exports. To attain this end, two conditions are necessary. First of all, there must be physical connection between the rail and water carriers. The railroad car must be brought alongside the pier where the river boat lies, just as it comes alongside the pier where the seaship lies. Lack of this connection means the expense of draying between the railroad's freight station and the river craft and this expense is usually sufficient to induce freight to stay in the car until it reaches its destination at the seaboard.

Equally important with this physical connection is the rate policy of the railroads, expressing a willingness to, or a determination not to, coöperate with the river and thus allow a large territory to enjoy the advantages of cheap rates to the sea. The railroads may not only refuse to prorate with the waterway as they do with other railroads, but they may apply high rates for their acts of feeding and distributing for the river, while they put their lowest tariffs on services parallel to it. Then, no matter how perfect the river port's contact between water and rail, it will be and remain cheaper for all inland points not on the waterway to communicate with the seaboard by rail direct rather than by transshipment into a barge at a river port. The railroad rate policy determines how large shall be the hinterland that the river is allowed to serve. Many a product remains on the farm, in the mine or unmanufactured because it is refused sufficiently cheap transportation to the sea. Many a needed

THE PORT OF HAMBURG

foreign product cannot bear the cost of rail transportation inland.

The merchants of a great seaport are not mere impassive middlemen through whose hands pass in automatic flow the surplus products and necessary imports of the country back of them. They are constantly finding new markets for the farms and industries inland and are constantly finding new sources of supply for foreign foodstuffs and foreign raw materials that are needed by the domestic population and factories. The seaport merchants are continually enlarging their transshipment trade to the coastal points and the nearby foreign countries whose foreign trade is too small for them to have direct lines of their own.

Lastly, a great seaport does not sit quietly by until enough indirect trade has grown up between it and a foreign land to make certain the success of a direct steamship line. Enterprising shipowners, with or without subsidies, establish coasting and oversea lines to develop trade. The merchants, exporters and importers of the port and the manufacturers inland unite to support the line and soon the trade is there. There are constantly being added to the port's equipment new coasting lines which, beside acting as collectors and distributors of foreign trade, carry goods in domestic commerce more cheaply than the railroads can. Similarly, the shipyards of the port may do their part to help its progress. In close coöperation with the steamship lines, they may be the pioneers in the designing and constructing of new ship types, so economical to operate that they increase the profits or the radius of action of the lines to which they belong, or so comfortable and so luxurious that they

INTRODUCTION

attract and hold the lucrative passenger and emigrant trade.

The seaport in this broader sense is the most important part in the mechanism of foreign trade. It is more than an interesting device; it is a living organism, or rather an essential part of that organism which we call the state, and has a vital function to perform. Its function is to call into life and handle the streams of foreign commerce and coastwise trade, to find for farms, mines and factories the markets and sources of foreign supply that they need, to organize and develop coastwise domestic trade. It is a function that is of increasing importance as the commercial bonds uniting nations become closer, the national and international specialization of production more complete, the volume of international exchange greater and the competition on the world's markets more severe.

Considered from this point of view it is apparent that the equipment and working of the leading seaports are a matter of great economic interest as bearing on the future of the nations they represent. Such is the justification for a study of the port of Hamburg. It has been among the primary factors contributing to the swift rise of Germany, since her rebirth in 1871, to a position of great economic power; it is one of the most valuable assets of the Empire today.

Hamburg has a perfect technical equipment, permitting the most expeditious, the smoothest possible passage of freight between sea and inland carriers. It has a network of waterways at its command and a railway system converging upon it from all parts of central Europe. Steamship lines give it probably a larger number of direct connections with foreign countries than any other port

THE PORT OF HAMBURG

enjoys. Moreover, this equipment is utilized as it should be. Port and channel dues are low. There is a heavy use of the Elbe inland and good coöperation with it on the part of the railways. The latter give their lowest tariffs to aid Hamburg against its foreign rivals or to further German exportation. Finally, the methods that Hamburg has pursued are most instructive for us because our ports are most like Hamburg.

This is evident when the type of harbor which Hamburg represents is considered. The other great north European ports beside Hamburg—London, Liverpool and Antwerp—are all, with the exception of a portion of the harbor at Antwerp, ports with closed docks. The difference between high and low tide at these ports is fifteen to thirty feet, such a difference that the construction of simple piers projecting out into the river was impracticable. Every tide would drop a foot of mud into the slips between such piers; no amount of dredging could keep them clear. A vessel rising and falling twenty feet with the tide, twice each day, would be most embarrassing to make fast, to load and unload. Moreover, the erection of deep-founded quay walls, deep enough to afford the vessels at their berths sufficient depth of water at low tide, high enough to carry pier sheds well above the high water mark—this would entail a prohibitive cost, though Antwerp has assumed a similar cost in the construction of a straight river quay of this sort three and one half miles long, alongside which many of the liners lie. The rest of the harbor at Antwerp and the entire harbor at both London and Liverpool, of which we hear so much, are pure closed-dock harbors. The principle of their construction and operation is one which, happily, has no

INTRODUCTION

practical significance for us. Their docks are entered through locks which at best are open only a few hours at high tide. That means that a ship may have to wait hours after entering the port before she can go to her berth or may have to wait hours after she has finished loading before she can leave her berth. This was a slight hindrance to sailing ships which sailed on no schedule and were so long at sea, so long loading and unloading, that a delay of a few hours in port was immaterial. But our liners today are on hard and fast schedules. A delay in getting to their berths is as serious as a slackening of speed at sea. Whoever sees the big Cunarders lying anchored in the Mersey, waiting to go to their berths, may admire the skill with which natural difficulties were overcome but cannot envy the English their dock system.

Fortunately, nothing of the sort is necessary in America. New York, for instance, has a difference of six feet between high and low water, Boston of nine and one half feet. Therefore we can use the same system that Hamburg, with her difference of six and one half feet between high and low tide, uses: piers and slips opening directly into the river. In New York, where the river is wide, piers were built out into it; in Hamburg, where the river is narrow—Hamburg is eighty-five miles from the sea—slips had to be cut into the land, leaving solid piers projecting. In the case of Hamburg, as here in America, it is possible to have open basin harbors whose piers can be approached at all tides. Then, if the channel is so dredged that the largest vessels can come up even at low tide, delay to the liners in reaching their berths has been abolished.

It is fully as important to shorten the unproductive

THE PORT OF HAMBURG

stay of the steamer in port by expediting her loading and discharging by the use of freight-handling machinery. This is a respect in which Hamburg has led the way and one in which we have unfortunately followed the older English ports, with their use of human labor instead of cranes for handling cargo. Though labor is dearer in the United States than in Germany and hoisting machinery cheaper, though on the Great Lakes we have developed the most perfect machinery for dealing with bulk cargoes, we have been unwilling on the Lakes and in our seaports to follow the lead of Germany in adopting machinery for handling package freight.

Moreover, the great English ports were built up before the days of the railroads. When the latter came upon the scene, it was not possible to lay tracks through the city to the centrally located docks. As a result, these are without adequate railway connection. We have already seen the importance of this connection for all ports dependent on trade with their hinterland and not, like London, primarily interested in re-exportation by sea or in supplying the needs of a local population as large as that of the three Scandinavian monarchies combined. Hamburg is dependent on the country behind it, as our ports are. In American ports, at least for all extensions, there is the possibility of creating that smooth contact between water and rail carriers which distinguishes Hamburg.

Finally, Hamburg has a river like the Mississippi and uses it in a manner that could be transplanted directly to the Mississippi system. Much of the eulogy bestowed on foreign waterways by trippers returning from Europe is expended on canals and ditches in Holland and Belgium.

Hamburg freighter at a pier, being served on her land side by cranes, on her water side by barges and lighters.

INTRODUCTION

These canals, like the little used English canals, if set down in America would be soon closed by the competition of American railways. Better canals have already fallen into disuse here. It is doubtful whether even the great network of canals serving Antwerp, which carry 300 to 400-ton barges to all parts of Belgium and France, could meet our American rail rates. The expenses of transportation attendant on delay in locks and limited speed in canal levels are so great that the days of the cross-country canal—even if it is toll-free—appear to be numbered throughout the railroad world. The future of inland water transportation lies in the use of free rivers like the Mississippi, the Rhine and the Elbe, these rivers collecting and distributing for a large territory. That this form of transportation is best exemplified today by the Rhine and the Elbe cannot fail to be the judgment of those who study waterways.

On the course of the Elbe, notably at Magdeburg, Torgau, Riesa and Dresden, are harbors which in technical excellence of construction, in providing for cheap transfer between river barge and car or dray, rival the works at Hamburg. Cranes, elevators, etc., facilitate this transfer exactly as at the seaport. On the river ply long, low, powerful side-wheel tugs, each having in its wake a train of barges of 600 to 800 tons average capacity. River transportation is in the hands of large, responsible companies which quote season rates, make deliveries as they promise to and, in conjunction with the steamship companies, give through bills of lading to all foreign lands. The steel construction of barges and tugs, the excellent condition of the river, the regulation of river traffic and the nature of river terminal facilities all work

THE PORT OF HAMBURG

to keep the insurance on water transportation low. By the use of the Elbe and the waterways connected with it, Hamburg extends her influence far into eastern Germany and Austria and brings to herself shipments that would otherwise go via Stettin or Trieste. It is the Elbe that has given Hamburg the supremacy over Bremen in Germany, for Bremen had only the short Weser to aid her—and the Weser goes nowhere and connects with nothing. To her rational use of the waterways at her disposal, more than to any other one cause, Hamburg owes her position today.

Hamburg has a superb equipment, partly the gift of nature, partly of her own creation. She uses this equipment in the most scientific and efficient manner to enhance her own prosperity but still more to further the development of the country she serves. The Germany of today is unthinkable without Hamburg, which is the symbol of German persistence, thoroughness, care of details, appreciation of opportunity and nice adaptation of the means to the end in view. The equipment of the port and the use of that equipment have been made under conditions similar to our own. Therefore a study of the port of Hamburg has more than the theoretical interest that attaches to a consideration of the construction and operation of any perfect thing. It has the practical interest that follows those achievements which show us the way to the removal of our own imperfections.

CHAPTER I

THE DEVELOPMENT OF HAMBURG'S HINTERLAND

UNTIL a short time ago Hamburg was primarily a transshipment harbor and an entrepôt for the countries on the Baltic Sea. Such was the nature of the Hanseatic seaports, Lübeck, Hamburg and Bremen. The Hansa was a loose league of North German cities whose merchants united to maintain common depots in foreign countries for collecting and distributing goods. The principal foreign depots were in London (the "Steelyard"), Bruges, Bergen and Novgorod. The Hanseatics controlled and carried the trade between Germany, the countries bordering on the Baltic and western Europe. They were able to monopolize this trade because for centuries the perilous voyage around the peninsula of Jutland was feared by merchant ships and because, when the sailors of the Hansa's rivals dared make the voyage, the Hansa had for a time power to close the Baltic to rival ships. This forced a large part of the trade from the North to the Baltic Sea to pass from Hamburg to Lübeck by an inland waterway and made Lübeck, the Baltic terminal of this waterway, the leader of the Hanseatic League and the natural entrepôt of the Baltic.

It was a considerable traffic, this between the wild East and the more cultured and developed West of Europe. From the fourteenth century on, Bruges, Antwerp, Amsterdam and London became each in turn the commercial center of northern Europe. The manufacture of woolen cloth grew up in Flanders and was later trans-

THE PORT OF HAMBURG

planted to England. These countries needed from eastern Europe grain, raw wool and wood for shipbuilding. Germany and the Baltic countries were purveyors of foodstuffs and raw materials to the more or less industrial northwestern Europe. French salt and wine, English and Flemish cloth and German knives were exchanged for wax, furs, Swedish ores, naval stores and other raw materials of the north and east, as well as Saxon and Silesian linen. Later, herring became the staple of the Baltic trade. So thoroughly was this commerce in the hands of the Hansa that the Scandinavian vikings, whose descendants play the second rôle in the European coasting trade today—after England—were driven from the sea. The fall of the power of the Hansa came when out of the political chaos in Europe nations arose like Holland and England with a naval power that they could use to further the interests of their own merchant marine. Against them the Hanseatics could not maintain their privileges.

It was this fact, not the discovery of America and the circumnavigation of Africa, which cost the Hansa towns their prosperity. Though the new discoveries brought into life a trade that was to dwarf that between East and West Europe, yet the latter trade remained. But it was no longer monopolized by the Hanseatic League. England led the way in sending her ships direct to Baltic ports without transshipment in Hamburg. The chief support of commerce from the Baltic, the herring fisheries, was removed when the herring changed his residence from the Baltic to the North Sea, there to become a bone of contention between the English and the Dutch. Though the monopoly of the Hanseatics was broken for all time,

HAMBURG'S HINTERLAND

traces remain of their power today. A striking example of this is the position in the Baltic trade which Lübeck still holds. Certain wares from London to St. Petersburg still go via Lübeck.[a] Hamburg, as we shall see, has today a heavy transshipment trade in Baltic countries in oversea goods, such as coffee, rice, tobacco, cacao and maize. The need for such a trade arises from no monopoly but from the fact that the Baltic countries have not sufficient trade with the oversea sources of these products to justify direct steamship lines thither. That Hamburg gets this transshipment trade is due largely to the fact that she has maintained through the centuries her ancient Hanseatic connections with the north and the east. Another factor that tends to make Hamburg the Baltic entrepôt is that it is the farthest east of the great north European ports. Other things being equal, it is a law in ocean transportation that the transshipment trade prefers to attach itself to that port which allows goods the longest use of the cheaper ocean liner, before transshipment into the dearer coasting vessel or "short trader."

Though Hamburg, like the other Hanseatics, felt the loss of the monopoly of trade between East and West Europe, she was in a better position than Lübeck to take part in the new commerce with Asia and America. As regards this commerce, Lübeck was in a cul-de-sac, while Hamburg faced toward the Atlantic Ocean where the new trade routes lay. But Hamburg was for centuries destined to a subordinate rôle in this new commerce which was soon to preponderate over all others. Portugal, Spain, France, Holland and England jealously guarded

---

[a] Schäfer: Article "Hansa" in Handwörterbuch der Staatswissenschaften, Jena, 1892.

the trade with their oversea colonies—and they owned the oversea world. Hamburg had to be satisfied with such crumbs as fell from the richly laden tables of the merchants of Lisbon, Cadiz, Amsterdam and London.

It was by acting as middleman between these rich emporia (particularly London) on the one hand, and Russia, the Scandinavian countries and Germany on the other, that Hamburg supported herself through the centuries that elapsed between her old supremacy and her new. It was Hamburg's duty to be on good terms with the countries having oversea colonies, in order to be admitted to the European carrying trade on at least an equal footing with all rivals. In rare cases Hamburg ships were even allowed to join the convoyed fleets that plied between European ports and the colonies. The demands on Hamburg's diplomacy were often great and her diplomacy had no naval power to back it. In 1670 the tattered German Empire had no ships, Hamburg had two convoys which escorted her merchantmen to meet the returning colonial fleets. At the same time Holland had 91 warships with crews totaling 23,500 men, England had 173 warships with 43,000 men aboard. An instance illustrates the difficulties which the Hamburg senators had to face. In 1666 the Dutch Admiral Brederode sailed up the Elbe, destroyed one Hamburg and several English ships. England threatened reprisals if Hamburg did not pay indemnity for the losses which England had sustained. Hamburg saw herself forced to give in.[a]

As Spain's power and her hold on her American colonies weakened, Hamburg did a thriving smuggling trade with South America and the West Indies, staples being

[a] R. Ehrenberg: Hamburgs Handel und Schiffahrt.

HAMBURG'S HINTERLAND

coffee, sugar, Brazilian pepper, indigo and dye woods. Hamburg plantations, which still endure, were established in Central America and the West Indies. In 1789 there were 5,000 men employed in the sugar refineries of Hamburg, which supplied the kingdoms of the North with sugar and manufactured tobacco.[a] A similar smuggling trade was a source of rich profit to Hamburg merchants during the American Revolution. The new United States of America were the first great foreign field open to Hamburg. When Mexico and the South American dependencies of Portugal and Spain revolted, Hamburg merchants exclaimed joyfully, "Hamburg has colonies at last." There was one great hiatus in this development: the years 1806-14, when the French occupied Hamburg, while Napoleon by his Berlin Decree and England by her retaliatory measure of blockading the Elbe shut off the port's communication with the outside world.

Apart from this period the first half of the nineteenth century was a time of steady progress. Those oversea colonies which had not revolted were gradually given freedom to trade direct with the European countries; the old policy of repression had not had happy results. More important still, there was growing up behind Hamburg an industrial hinterland, which was to become her sure basis of prosperity. Ever since the Peace of Westphalia, in 1648, Germany had been split up among a hundred princelings, each with his own customs line, his own road and river tolls. There was no possibility of anything but local trade. But in the twenties the German Customs Union was formed under the leadership of Prussia, eager to make a market for her promising industries. The

[a] Plath, page 99.

THE PORT OF HAMBURG

Customs Union spread until by the middle of the century it included the greater part of Germany. As customs barriers fell, traffic tolls—there was a toll to pay every few miles on the Elbe—were cut down, the Elbe being declared toll-free in 1870. Beginning with the forties railroads were built, for the first time bringing within reach of Hamburg towns and districts not on the Elbe or its connecting waterways.[a]

Important as were the building of railroads and the work of the Customs Union in making the interior of the country accessible, leveling tariff walls and giving a field of expansion for German industries, it was not until after the formation of the Empire in 1871 that Germany was sufficiently interested in foreign trade to become the basis of a new prosperity in Hamburg. As late as 1869, 60 per cent of the register tonnage of ships that entered Hamburg came from England, only 15 per cent from the rest of Europe.[b] That meant that Hamburg was still playing middleman for London. Napoleon characterized Hamburg when he exclaimed, "*Ne me parlez pas de cette ville anglaise*"; until recent years there were Hamburgers

[a] At the beginning of the nineteenth century the dry land behind Hamburg was thought of less as a source of commerce than as a source of marauding expeditions. The marshes that isolated Hamburg and Antwerp from the land were looked upon as advantages in that they made the cities inaccessible to robbers. In 1804 Hamburg on either side of the Elbe was surrounded by high walls, their continuation across the Elbe being represented by booms, obstructions preventing the passing of boats. From sunset to sunrise the gates were closed, the booms swung across the river and the city locked up hard and fast for the night. From January 1 to 12, according to the official almanac, this period from sunset to sunrise lasted from 4.15 p.m. to 8 a.m. Communication inland was solely by the Elbe and its tributaries.

[b] Wiendenfeld, Hamburg, page 10.

who were proud that they had never been in Berlin. In 1871, when Hamburg and Bremen entered the German Empire, they stipulated that they should remain outside the Customs Union. Their transshipment trade still prevailed.

After the formation of the Empire a new order of things dawned. Five billion francs of indemnity from the defeated French poured into the country. There were imported the industrial processes that England had developed: the automatic loom, the blast furnace for pig iron, the Bessemer converter for steel. In 1879, Germany enacted a protective tariff to develop her industries and save her threatened agriculture, against which the grain of the American prairies was flowing in. Both manufactures and agriculture responded to the opportunity. There was occurring in the economic world an interlocal and international specialization of labor that resulted in an enormous exchange of goods by sea. Cheapened production and cheap transportation made possible the exchange of goods that had never moved from the spot before. It was this change in Germany's economic life that decided Hamburg to accede to the importunities of Bismarck and enter the Customs Union. Yet so strongly did the belief in the old transshipment trade persist that the greater part of the port was fenced off and set apart, to remain a Free Port, outside the Customs Union, just as the whole city had been outside of the Union before.

The protective duty on meat has encouraged cattle-raising; that meant imported cottonseed cakes from America, maize from America and the Argentine to fatten stall-fed cattle. The protective duty on grain has encouraged the fertilization of agricultural lands; that

THE PORT OF HAMBURG

brought with it imports of Chili saltpetre and guano and of phosphates from Florida. A similar demand abroad led to the exportation of potash from the inexhaustible German fields on the Elbe. The perfection of the process of manufacturing sugar from the sugar beet resulted in Germany becoming the world's first producer and a heavy exporter of the article.[a] Finally, the change during the nineteenth century from the use of wood to the use of coal for industrial and domestic firing led to a movement of coal that has become the backbone of commerce. In 1907, 40 per cent of the traffic tonnage on the German railways consisted of coal and coke; four and one half million tons imported coal constituted one third of the tonnage of the imports of Hamburg.

Yet Germany could not be made capable of supporting the enormous increase in her population, which rose from forty million in 1870 to sixty-five million in 1910—at the rate of 900,000 per year the last ten years. Nor is this surplus considerably reduced by emigration, as formerly.[b]

[a] Production of beet sugar, year 1906-07:

Germany,	2,235,000 tons
Austro-Hungary,	1,334,000 tons
France,	756,000 tons
Belgium,	283,000 tons
Holland,	181,000 tons
Russia,	1,470,000 tons
Other Lands	445,000 tons

(Hamburgs Handel, 1907.)

Throughout this book, "ton" means a metric ton, 2,204 English pounds.

[b] Yet in the years 1881-90, the years of the heaviest emigration, only 24 per cent of the natural increase of population were carried off. Since 1895 there is an excess of immigration. In the years 1871-95, the excess of emigration over immigration was two and one

HAMBURG'S HINTERLAND

German emigration dropped from such figures as 210,000 in 1881 and 250,000 in 1882 to 32,000 in 1895 and has never since passed 50,000. The acreage of German soil sown with grain increases but slightly. The yields of wheat, barley, rye and hay have not increased in the last decade in spite of higher protective duties. The production of grain in Germany seems to have reached the saturation point and an increasingly large portion of the yearly consumption of grain must come from America, Argentine, Roumania and Russia. The figures for the national production and the importation of wheat indicate this.

SOURCES OF GERMANY'S WHEAT SUPPLY.[a]

	Raised	Imported	Value of Imports Million Marks
1899	3,847,447 tons	1,370,051 tons	180
1903	3,555,064 tons	2,124,643 tons	253
1907	3,479,324 tons	2,634,889 tons	385

To pay for this great increase in imported foodstuffs, Germany has had to export manufactured goods. How successfully she has met this need is seen in the following table, indicating the increase of her exported manufactures since 1890.

EXPORTS AND IMPORTS OF MANUFACTURED ARTICLES.

	Million Marks	
	Exports	Imports
1890	2,148	981
1899	2,712	1,148
1907	4,638	1,392

half million persons; in the years 1895-1900, 94,000 more emigrants settled in Germany than left it. (Die Seeinteressen des deutschen Reichs, and Die Entwicklung der deutschen Seeinteressen.)

[a] Statistische Jahrbücher für das deutsche Reich.

THE PORT OF HAMBURG

Moreover, the raw materials for many of these manufactures come from oversea; for instance, the raw materials for cotton, woolen and silk goods. Germany's imports of raw cotton have increased within recent years as follows:[a]

IMPORTS OF RAW COTTON.

	Million Marks[b]
1901	296
1903	395
1907	551

Germany is fast becoming a typical industrial country; that is, a country engaged in importing the raw materials of industry and foodstuffs for supporting a population that pays for these imports by exporting its manufactured products. This is apparent from a consideration of the following table, which shows the enormous growth of foreign trade since 1890 and the principal items of export and import in the year 1907.

GROWTH OF GERMANY'S EXPORTS AND IMPORTS.

	Million Marks	
	Imports	Exports
1890	2,860	2,946
1899	5,786	4,364
1907	8,747	6,845

[a] Statistische Jahrbücher für das deutsche Reich. In the period 1894-1904 Germany's foreign trade increased 66 per cent, England's 38 per cent, the United States' 59 per cent, France's 28 per cent, Russia's 23 per cent.

[b] A mark is 23.8 cents. In rough reckoning it is convenient to consider it 25 cents. There are 100 pfennigs in a mark. A pfennig is roughly one fourth cent.

HAMBURG'S HINTERLAND

LEADING ITEMS.

IMPORTS.		EXPORTS.	
Cotton,	551	Cotton Goods,	432
Wool,	394	Machines,	412
Wheat,	385	Chemical Products,	350
Barley,	282	Coal,	280
Coal,	242	Woolen Goods,	286
Copper,	240	Silk Goods,	204
		Sugar,	193

An analysis of the population of Germany by occupations according to the censuses of 1882, 1895 and 1907, respectively, shows that the surplus of population is being cared for primarily by industrial employment. Only the three principal occupations—industry, commerce and agriculture—need be considered. The percentages are percentages of the total population.[a]

OCCUPATIONS OF GERMAN POPULATION.

Employed in	1882	1895	1907
Agriculture,	19,225,455	18,501,307	17,681,176
Manufacturing,	16,058,080	20,253,241	26,386,537
Commerce,	4,531,080	5,966,846	8,278,239

PERCENTAGES.

	1882	1895	1907
Agriculture,	42.5	36.2	32.7
Manufacturing,	35.5	36.1	37.2
Commerce,	10.	10.2	11.5

Germany is a continental state, centrally situated in Europe, with long terrestial borders and a comparatively short coast line. Yet an extraordinarily large part of her foreign commerce is by sea. All of the leading items of her imports, given above, are from oversea, as are

[a] Statistische Jahrbücher für das deutsche Reich.

THE PORT OF HAMBURG

colonial wares such as coffee, tea, rice, tobacco. Similarly, her markets for manufactured goods are mostly across the water: either in such free trade lands as England, South America, India and China, or in the wealthy United States. Of Germany's 15,592,000,000 marks of foreign trade in 1907, 34.5 per cent[a] were with extra-European countries, 18 per cent more were with England and Scandinavian countries and hence carried on by sea. Moreover, a large commerce with European countries goes by sea: that with countries bordering on the Mediterranean and the Black Sea. Finally, a large portion of Germany's foreign trade is carried on via Rotterdam and Antwerp, which act as seaports for the industrial West Germany. Thus one third of the volume of Germany's oversea exports and imports appears in the statistics as destined for or coming from Holland and Belgium.[b] A German official publication, "Die Entwicklung der deutschen Seeinteressen," estimates that 70 per cent of Germany's foreign trade is by sea. The meaning becomes clear of that watchword which the Kaiser gave out at the dedication of the new harbor at Stettin: "*unsere Zukunft liegt auf dem Wasser.*"

The interest of Hamburg in this development is apparent when we consider that Hamburg's natural and immediate hinterland is the seat of the beet sugar industry, intensive farming and cattle-raising and that Germany's potash deposits lie on the banks of the Elbe, Hamburg's stream, and on the Saale, its tributary. On the Elbe and its connecting waterways are such manufacturing centers

[a] This percentage was 31.5 per cent in 1904, 28.2 per cent in 1898, 27.1 per cent in 1894.

[b] Die Entwicklung der deutschen Seeinteressen. Einleitung, VI.

A row of old warehouses on the Steckelhorn, one of Hamburg's ancient canals.

HAMBURG'S HINTERLAND

as Magdeburg and Dresden, Berlin and Breslau—and back of the latter the Silesian industrial district. The Saxon industrial district finds in Hamburg its natural outlet. The industries of Westphalia are brought within Hamburg's sphere of influence by favorable railway tariffs. The result has been that Hamburg's commerce by sea has increased even more rapidly than Germany's population and Germany's foreign trade; i.e., that Hamburg handles a continually larger portion of that foreign trade.

GROWTH OF GERMANY'S FOREIGN TRADE AND HAMBURG'S PART THEREIN.[a]

	German Population	German Foreign Trade	Hamburg's Seaward Trade [b]	Index Numbers		
		Marks	Marks	German Population	German Trade	Hamburg Trade
1880	45,095,000	5,805,100,000	1,700,100,000	100	100	100
1907	62,000,000	15,586,000,000	6,397,500,000	137	269	376

Exports and imports for Hamburg's hinterland now dwarf into insignificance the transshipment trade to Scandinavia, Russia and the Baltic provinces. Direct lines are employed to carry goods between Hamburg and oversea; the English middleman has been shaken off. The process of emancipation from the English middleman is graphically illustrated by the following table:

[a] Statistisches Jahrbuch, Hamburgs Handel und Schiffahrt, Die Seeinteressen des deutschen Reichs.

[b] This includes goods exchanged with German ports in the coasting trade. However, it is fair to assume that the proportion of the coasting trade in the total is the same for both years.

THE PORT OF HAMBURG

Source of Imports in Hamburg.[a]

	Million Marks		
	1871-80	1896	1907
Non-European Lands,	262	958	2,222
United Kingdom,	474	410	635
Rest of Europe,	139	345	1,355

This increase in imports direct from oversea lands is not for transshipment. It is for Germany's intensive cultivation of her soil that the immense quantities of Chilean nitrate arrive in Hamburg; the shiploads of maize from Baltimore are to become German beef. The colliers that ply between Newcastle and Hamburg are bringing fuel for the industries of Hamburg and Berlin and bunker coal for half the German merchant marine. Millions of tons of grain, thousands of tons of coffee are brought to feed the workers that are to transform the bales of cotton, the Australian wool and East Indian jute, the iron pyrites from Spain, the American and African woods that are piled high on the quays and in the pier sheds of Hamburg.

And so with the exports that we see leaving Hamburg. The sacks of sugar that are being loaded into the ships came from the scientifically cultivated fields of Silesia, Prussian Saxony and Bohemia; the barrels of alcohol are from the distilleries of Brandenburg and Pomerania. The heavy iron work and the machines, shipments of cotton and woolen wares, cases of leather goods and glassware, the miscellany of chemical products, are from German workshops. The potash, salts and coal that ballast the outgoing ships come from German soil.

[a] Statistisches Jahrbuch, 1907, and Die Seeinteressen des deutschen Reichs.

HAMBURG'S HINTERLAND

The old Hanseatic seaport, turned exclusively toward the sea, has become a new one turned toward Germany—become a part of Germany and dependent on it. When an ambitious Frenchman inquired in Hamburg how Nantes could follow the example of Hamburg, he received the reply: "You must commence by transforming at least Orleans, Tours, Blois, Saumur and Angers into cities of from 200,000 to 300,000 inhabitants and cover with factories the country roundabout. Then perhaps Nantes will be able to follow the example of Hamburg."[a] The new prosperity of Hamburg has its basis in the prosperity of the new Germany.

[a] Paul de Rousiers, page 177.

CHAPTER II

THE CHANNEL TO THE SEA

THE development of Hamburg's hinterland has been sketched in the previous chapter. In the industrial development of Germany there was created the need for seaports to handle its increasingly large commerce with oversea. We have now to investigate the manner in which Hamburg, favored by an exceptional location, took advantage of its opportunity and became the greatest of European seaports. Next to the provision of modern harbor and terminal facilities, its most pressing need was the creation of a channel to the sea, adapted to the conditions of today.

Of increasing importance to modern harbors are their channels to the sea. North European seaports were originally set at the point where river and ocean navigation met: the ocean-going vessels proceeded as far inland as they could. How far inland they could penetrate is indicated by the fact that one of the great Hanseatic seaports was Cologne on the Rhine, 210 miles up the river from the Hook of Holland.[a] The channel from the sea to Cologne is today only nine feet, after the expenditure of millions of marks. In mediæval times it could not have been six feet. Yet a channel of less than

[a] Columbus' three ships had the following dimensions: "Santa Maria," 300 register tons; "Pinta," 100 register tons; "Nina," 30 register tons. Their carrying capacity was approximately double their register tonnage. The "Santa Maria" was about as capacious as a small Elbe river barge. The "Pinta" was the size of a large harbor lighter, the "Nina" the size of a small one.

THE CHANNEL TO THE SEA

six feet sufficed to make Cologne a seaport like Hamburg and Lübeck and Bremen. It was to be accessible to seaships of such draught that Hamburg was located where it lies, eighty-five miles up the Elbe, Bremen seventy-five miles up the Weser, Antwerp fifty-nine miles up the Scheldt. The enormous expenditures necessary to adapt these channels to modern requirements are apparent when we consider into what the seaship of six feet draught has developed. The "Kaiserin Auguste Victoria," the largest steamer entering Hamburg, draws nearly thirty-three feet fully loaded, the "Mauretania" and "Lusitania" thirty-seven feet. The Cunard, White Star and Hamburg-American Lines are each building liners which, each of 45,000 to 50,000 register tons,[a] are to dwarf all previous creations.

Most of this increase in the tonnage and draught of vessels, and therefore in the depth of channel necessary to accommodate them, has occurred in the past century. In 1847 the Hamburg-American Line began its career with a new sailing ship, the "Deutschland," the pride of Hamburg. She was of 717 register tons. The "Kaiserin Auguste Victoria" today has a gross register tonnage of 25,000. Correspondingly, Hamburg was satisfied in the forties with a channel of six and one half feet at low,

[a] The British register ton is an arbitrary unit of measurement for merchant ships; it represents 100 cubic feet of space. A ship's total space available for sheltering merchandise, persons or machinery is expressed in register tons, gross. Its net register tonnage is this total minus the space occupied by machinery, officer and crew quarters, instrument rooms, etc. The net register tonnage of the ship represents its capacity for carrying passengers and goods, including coal and stores. The relationship of gross to net tonnage is in general 5:3 and in a freighter a net register ton means, on the average, capacity for 1¼ to 2 tons of freight.

THE PORT OF HAMBURG

thirteen feet at high water. The normal sailing vessel of the forties was of 200 to 300 register tons, or about 450 tons carrying capacity. Barges of 1,000 tons capacity are now towed up the Elbe 500 miles, to Prague, in Bohemia.

Then came that remarkable change in ocean commerce which is characterized by the appearance, for the first time, of great quantities of bulk goods of low specific value. Improved means of transportation—the steamship on the sea and the railroad on land—made it possible to move these goods from land to land. The physical limitations to the size of wooden sailing ships did not affect iron and steel steamers. They can be built indefinitely large, and, provided there is a sufficiently large stream of traffic, they are profitable to operate directly in proportion to their size.[a]

The Elbe valley from the sea to a point above Hamburg was originally a bay of the North Sea. When Hamburg was founded, the valley had mudded up and was an uninhabitable bog, through which the river wandered, in many constantly shifting channels, to the sea. The principal work before the nineteenth century was, by the erection of dikes, to reduce the number of these arms. Until the middle of the nineteenth century, the lack of depth of the channel was not felt to be a disadvantage.

[a] It was the Germans who first recognized the greater profitableness of the larger steamers. In 1901, although the British merchant marine was five and one half times larger than the German, England possessed only 28 steamers over 10,000 tons gross, while Germany had 24, America 6. (Report of the Royal Commission on the Port of London, 1902, page 26.) Today the Hamburg-American Line alone has 11 steamers of over 13,000 gross register tons each, America 3. (Hamburg als Schiffahrts- und Industrieplatz, Mai, 1910; and Report of the U. S. Commissioner of Navigation, 1909.)

THE CHANNEL TO THE SEA

Besides reducing the number of arms of the river and attempting to give the main channel a fixed location, improvements included buoying off the channel and maintaining a lighthouse at its mouth. In 1246 the Archbishop of Bremen gave Hamburg half the island of Neuwerk, near the present site of Cuxhaven, on condition that a lighthouse should be maintained there. One of wooden construction, with stone foundation, was built, and an open fire was kept burning on an upper platform. It is the oldest lighthouse in the world still in use.

If steamers had not supplanted sailing ships it would probably have sufficed to create a channel up which the sailers could have come at high tides. Even the packet sailers operated on only a nominal schedule. They arrived on time "wind and weather permitting." The time they took up in the journey was so great that a delay of half a day waiting for high tide would not have seriously inconvenienced them. But steamship schedules are published a season in advance; incoming and outgoing liners connect with special trains. Some of the liners of today cost ten to twelve million marks. The profit from such vessels depends on the number of trips they can make in a year, and this number must not—if it can be avoided—be reduced by having to wait for high water to come up to the port. The ideal is for the channel to be so deep that all vessels using it can go up the river at low water. The next possibility is to have the channel so deep that at low water most of the ships can go up, while the very largest wait for higher stages of water. The third possibility is to give up hope of having the largest steamers reach the port and, in view of this, creating for them a sub-port nearer the mouth of the river.

THE PORT OF HAMBURG

This last course is a very serious matter. Ocean rates are no cheaper to the sub-port than to the main port higher up, while the railway rates inland, which go by mileage, are considerably higher for the more distant sub-port. Moreover, unless the costly warehouses of the mother port are to be duplicated downstream, all vessels discharging warehouse goods at the sub-port must send them by lighter or rail to the warehouses upstream. London has chosen this course. Tillbury docks are the sub-port; warehouse goods coming, for instance, by the Atlantic Transport Line or the Peninsular and Oriental are lightered (or railed) to the warehouses in London, Tillbury docks being about twenty-five miles below London Bridge. Bremen has had to pursue the same course and construct in Bremerhaven, near the mouth of the Weser, a duplicate of its harbor at Bremen, excepting for warehouses, in order to accommodate the North Atlantic liners of the North German Lloyd. The ideal of having a channel up which all ships could come at all times is not attained either by Hamburg or its competitors in northern Europe: Bremen, Amsterdam, Rotterdam, Antwerp, Havre, London and Liverpool. Our American North Atlantic ports are preëminent in that they are, or soon will be, accessible at low tide to all ships that care to enter them: New York with a channel of forty feet at low water, Boston with thirty feet (soon thirty-five), Philadelphia with thirty feet (soon thirty-five), Baltimore with thirty feet.

Liverpool, Antwerp, Rotterdam, Havre and Hamburg have had to be satisfied with a partial attainment of the ideal. In all of them, large ships can enter only at the higher stages of water. No beneficent national govern-

THE CHANNEL TO THE SEA

ment has dug Hamburg's channel for her; the little city-state has done her own dredging, as Bremen has.[a] In Germany, river and harbor improvements fall to the states in whose territories they lie and are no concern of the Empire. The Elbe below Hamburg flows between the Prussian provinces of Hanover and Schleswig-Holstein. Under treaties, first with the kingdoms of Hanover and Denmark, later with Prussia,[b] after she had annexed Hannover and Schleswig-Holstein, Hamburg has done all the work on the river.

In the forties Hamburg was possessed of a channel to the sea of six and one half feet at low water, thirteen feet at high.[c] In order to see what Hamburg has had to strive after, it is well to consider the main types of ocean-going steamship that have developed for her. Tramp steamers are not usually of over twenty-one to twenty-four feet draught. The liners are divisible, as far as Hamburg is concerned, into three classes: the steamers in the European service, the steamers in the Suez service to the Far East, the liners in the transatlantic service. The European liners have a maximum draught of about nineteen and one half feet, determined by the depth of the numerous harbor-channels they must enter. The draught of the Suez liners is determined by the regulations of the officials of the Suez Canal. Up to 1902 the permissible draught in this service was twenty-three feet (seven meters[d]): it was then raised

[a] Yet J. Paul Goode states in his report to the Chicago Harbor Commission, referring to Hamburg: "The German Government takes care of the channel." (Report of the Commission, page 134.)

[b] The relations between Hamburg and Prussia in this respect are regulated by the Köhlbrand Treaty.

[c] Wiedenfeld: Die Welthäfen, page 52.

[d] A meter is 3.28 feet, roughly 3¼ feet.

THE PORT OF HAMBURG

to twenty-six and one fourth feet. Liners of the transatlantic service to New York draw, fully loaded, up to 32.8 feet (the "Kaiserin Auguste Victoria").[a]

Hamburg has striven and still strives to keep a channel that will let the European liners come up to the harbor at all stages of water, the Suez liners excepting at low water: the transatlantic liners are to be able to get up at high water without lightering, which is a costly performance.[b] For years the large liners from New York, in order to reach Hamburg even at high water, had to lighter a part of their cargo at Brunshausen, twenty miles below the harbor, at the lower end of a bar which has occasioned most of the dredging. Outbound ships had to complete their cargoes from lighters after they had passed the bar. It was the costliness of these operations that induced the Hamburg-American Line to consider despatching its New York liners from Cuxhaven, at the mouth of the river, eighty-five miles down stream. The

[a] It must not be thought that vessels can load to utilize the full channel depth. When moving at any speed, a vessel "settles" below her stationary loaded draught. Moreover, when the depth of channel is too completely utilized, the vessel "feels" the bottom and answers her helm badly. One meter ($3\frac{1}{4}$ feet) is a small margin to leave between keel and bed of the river; 31.16 feet is the maximum draught of a vessel entering the port of Hamburg. At present two steamers, the "Kaiserin Auguste Victoria" and the "Amerika," are not ordinarily brought up above Brunshausen without lightering at that point. Here they also complete their cargoes, outbound. These two ships of this class, on arriving from New York, can sometimes reach the port with depleted coal bunkers, but cannot clear with full bunkers and full cargo. (Barge Canal Report, pages 396, 416.)

[b] Rates for lighter service between Brunshausen and Hamburg are 1.80 marks per ton (42.84 cents) in summer; 2.50 marks (59.5 cents) per ton in winter, or more, if ice conditions are particularly unfavorable.

THE CHANNEL TO THE SEA

unfortunate results of the first year's occupancy of Cuxhaven and the prospect that the new regulation of the Elbe, begun in 1902, would make it possible for vessels of the greatest draught to reach Hamburg without lightering, caused Cuxhaven to be abandoned by the liners except as a port of call.

Fortunately, at the same time that steamships of rapidly increasing size demanded a deeper channel, the steam dredge was invented, which alone made a deeper channel possible. In addition to dredging, the channel has been deepened by the construction of fascine-work moles, which run out from either bank of the river, slanting down stream, their backs just below high water. The distance between the heads of opposite moles is the desired width of the channel. The river bed fills up between consecutive moles and the current, confined to the narrower channel, has sufficient force to scour it and prevent it from sanding up. An elaborate system of buoys on the water, landmarks and lights on the land, is maintained, so that ships can come up the river by night as well as by day. Ice breakers have kept the river open for navigation all winter, since 1876. In 1800 skating and racing on the lower Elbe were annual diversions of the Hamburgers; the Elbe froze so hard that vessels were "loaden and unloaden on the ice."[a]

In the summer of 1910 the Hamburg officials could boast that all large ships entering the Elbe had been able to come up to the harbor and that not a vessel had been compelled to lighter during the year. That is, accessi-

[a] The Picture of Hamburg. Navigation on the lower Elbe was closed, because of ice, for sixty-one days in 1847, eighty-one days in 1855, fifty-three days in 1871. (De Rousiers, page 209.)

THE PORT OF HAMBURG

bility, at high water and without lightering, had been attained for the largest boats in the transatlantic service. But for one consideration accessibility for ships of the Suez service at all but the low stages of water would have been attained; twenty-six and one fourth feet was the low water depth of the channel. The reason why this was not accomplished is one that indicated how endless is the task of channel improvement. On January 1, 1910, the draught permissible to steamers using the Suez Canal had been raised to twenty-nine and one fourth feet (nine meters).

So at present Hamburg has a channel to the sea, eighty-five miles long, of 650 feet width (200 meters), twenty-six and one fourth feet depth at low water (eight meters), 32.8 feet at high water (ten meters). Two considerations make it certain that the work will proceed. One is that one meter more is needed to bring the deepest-going Suez boats up at all but low stages of water, and this is part of Hamburg's policy. The other reason is: the Hamburg-American Line is building in the "Imperator," 50,000 tons gross, a liner twice as large as the present largest ship entering the Elbe, the "Kaiserin Auguste Victoria," 25,000 tons gross. At present the "Imperator" could not get up to Hamburg at all. It is improbable that this new colossus is being constructed without an understanding with the Hamburg state that there will be a channel ready for her when she is finished.[a]

[a] As this goes to press, State Engineer Bubendey writes from Hamburg: "We are now working on a further deepening of the Elbe from Hamburg to the sea. A low water channel depth of nine meters is being created; it will be ten meters later. These will correspond to high water depths of eleven and twelve meters, respectively."

The "Kaiserin Auguste Victoria" (25,000 tons), the largest ship entering Hamburg.

THE CHANNEL TO THE SEA

Today the most important element in the competition of the ports of northwestern Europe is their relative distances from the common hinterland. Hamburg has been, and still is, ready to spend large sums to bring her ships eighty-five miles further inland. Expenditures on the channel from 1850 to 1906 totaled sixty-two million marks.[a]

Up to 1861 there was a toll on all ships using the Elbe, collected by the government of Hanover at Stade, between Hamburg and the sea. In 1861 this toll was bought off, through the payment to Hanover, by a number of maritime nations, of a sum approximating $750,000. In a similar manner the Sund toll, levied by Denmark on ships entering the Baltic, had been bought off a few years before; in 1862 the Dutch were similarly compensated for removing the toll on navigation passing up the Scheldt to Antwerp, the commercial rival of their seaports. These were three remains of a vast and burdensome system of traffic tolls which stood in no relation to services rendered in aid of navigation.

Hamburg collected no formal due to repay her for her expenditures on the channel until the last expensive dredging in the years 1902-05. Since then she has collected a tonnage due on all seaships above a small minimum tonnage. It goes under the name of "buoy due" (Tonnengeld) and "declaration due" (Deklarationsgeld), and appeared at 3,305,000 marks in the estimate for 1906.[b] Vessels arriving from sea pay 12 pfennigs per cubic meter net: 8.1 cents per net register ton.[c] Vessels with cargoes

[a] Peters, part II., page 277.
[b] Richter, page 29.
[c] A register ton is 2.83 cubic meters.

of coal, cement and various other bulk goods, pay half the regular tonnage due, as do vessels arriving in ballast and departing with cargo. Ships that come and leave in ballast are exempt from tonnage dues; likewise ships built in Hamburg and returning from their maiden trips.[a] Small steamers and lighters coming direct from points on the German Rhine, as well as lighters that come from points on the Dortmund-Ems Canal, are also exempt. The purpose of this is to offer every inducement to the commerce of rich western Germany, which naturally gravitates to the nearer ports of Rotterdam and Antwerp.

It is doubtful whether the annual three and one third million marks of channel dues represent anything more than a nominal return, after maintenance is paid for, on the sixty-two million marks that have been spent on the channel since 1850. It is Hamburg's advantage that it is a state as well as a city, that it has no limit of bonded indebtedness imposed by a superior power, that it is not obliged to make any specific return on money invested in the harbor and channel. Hamburg proceeds on the principle that a tax on traffic is bad. The mere fact that a shipment is made via Hamburg is no evidence that its owner is able to pay a tax sufficiently high to make a full return for all facilities afforded him. When the unhindered flow of commerce has crystallized into wealth in the hands of merchants, forwarders, warehousers and shipowners, it is time to tax this wealth, primarily through a property and an income tax.

Stettin, Hamburg's rival on the Baltic, who dreams of becoming the third great German seaport and who is

[a] Barge Canal Report, Vol. I., page 397.

THE CHANNEL TO THE SEA

Hamburg's competitor for the foreign trade of Berlin and Silesia, is a Prussian seaport and has had its channel dug by the Prussian government, down the river Oder and through a shallow bay (Haff) to the sea. On April 27, 1909, a Stettin representative in the Prussian Diet arose to complain that the government's treatment of Stettin had made it impossible for that port to compete with Hamburg. "To be sure, the government has dredged a channel for us from Swinemünde to Stettin, but it has introduced a very unpopular channel due, levied on every ship coming up the Oder. A ship that has 270 marks in dues of this nature to pay in Hamburg, pays 1,102.50 marks in Stettin. In Hamburg they pay six pfennigs per cubic meter tonnage, in Stettin we pay twenty-eight pfennigs. Under such conditions Stettin can hardly hope to meet successfully the tremendous competition of Hamburg."

Von Breitenbach, the Prussian Minister of Public Works, was present and answered the complaint with arguments that show the advantages that Hamburg possesses in being a state. He said, in part: "The Prussian state carries a heavy burden because of this channel. After paying maintenance, the dues represent 1.17 per cent on the capital invested, and no more; 644,000 marks per year are lacking to constitute a proper return on the money we have spent. We have done all we can for Stettin. It is not allowable to compare Stettin with Hamburg and Lübeck. The latter are purely commercial states, which can devote their entire interest to fostering their commerce. The Prussian state has far wider interests and obligations. We cannot exert our entire influence in behalf of a single commercial port."

THE PORT OF HAMBURG

Hamburg, lying far up the Elbe, has seen itself compelled to spend large sums to allow the leviathan steamers of today to reach the port. It was willing to spend these sums rather than to consent to the creation of a sub-port for the larger ships, because it realized that all freight discharged at the sub-port would have heavier railway charges to bear. Not only are all vessels using the port able to reach it today; but provision is also being made for a channel that will suffice even the "Imperator" when she is finished. Hamburg, being an independent state, has had to dig the channel herself, but is free from the necessity of charging for its use dues that discourage traffic—an advantage that Hamburg enjoys over all its German rivals excepting Bremen.

CHAPTER III

PORT FACILITIES

IN VIEW of the present widespread agitation for improving the port facilities of our American harbors, the importance of this subject need not be emphasized. Port facilities mean provision for the proper contact between the ocean carrier and the coastwise vessel, between the ocean carrier and the railroad, and between the ocean carrier and inland waterway craft, if there be such. Warehouses, destined to shelter for more or less long periods goods sent through the port, must have proper connection with the inland, ocean and coastwise carriers. If the emigrant trade is sought, suitable accommodations for it must be furnished. Local exporting industries attach to the port an inbound and outbound traffic which nothing can take away from it. Other things being equal, that port will distance its competitors which provides the best, cheapest and most expeditious terminal, transshipment, warehousing, emigrant and industrial facilities. As in America channels to the sea are constructed for our ports by the national government, the lesson from Hamburg for our cities must be primarily in the matter of port facilities.

The original Hamburg harbor was on the river Alster, a small stream which flows through the city. As the wall and moat of the city were repeatedly pushed outward, the old moat became a canal, on whose banks the warehouses were erected which served Hamburg's transshipment trade. The small seaships penetrated the canals and

THE PORT OF HAMBURG

came directly to the warehouses. Not until the seventeenth century, when the Alster and the canals had become overcrowded, did the Elbe itself come into use as a harbor. It became more important when, after 1800, the larger sailing ships were no longer able to enter the canals.

Until 1866 the Hamburg harbor consisted of a stretch of river, with mooring posts driven into the river bed, to which the ships made fast. By human labor and the ship's tackle, they discharged into small lighters, which were poled or carried by the tide upstream to a hand crane on the bank or into one of the many canals on which the warehouses lay. The canals are not so numerous as they were, but there are still enough of them to make the stranger understand why Hamburg was called the Venice of the North. The warehouse wall rises directly from the water's edge, so that the lighter can lie close alongside. Above the door of the top story the arm of a hoist projects; it brings goods up from the lighter and they can be pulled in at the door of any floor. The warehouses still occupied—and there are many of them—operate their hoists electrically; they used to be wound up by hand. The canal frontage of the deep narrow building was a warehouse, the street frontage often the merchant's residence. Some of these canals are little changed and a trip through one in a row boat or a launch at high water—at low water they are almost dry—gives one a strong impression of having dropped in upon the fifteenth century.

When steamships came in, this method of discharging the cargo would no longer suffice. The steamship clamors for punctuality and speed in loading and discharging. Its profits depend on the number of voyages it makes in

PORT FACILITIES

the year. Cargoes became so huge and various that sorting them on the ship's deck for distribution into the lighters of numerous warehouses and into river barges was an endless task. To meet this difficulty large lighters were at first employed to act as floating piers. Into these the steamer dumped its burden; the goods were there sorted and then given over to the various small harbor lighters. But experience in English harbors had shown that quay walls with deep foundations, which allowed the ship to lie alongside the land and discharge into freight sheds, considerably hastened and cheapened the discharge of a cargo. Moreover, the rapid extension of railway transportation brought with it the need for direct contact between the ship and the railroad car.

These considerations made Hamburg decide to provide opportunity for its liners to come directly to land. English engineers were called upon to prepare plans for the construction of a modern harbor. In total disregard of the difference in conditions between London or Liverpool and Hamburg, they recommended closed docks with lock gates, like those of the English ports. In spite of the opposition of State Engineer Dalmann, the construction of such a dock was begun, but the superfluity of the entrance lock was seen before the construction was finished and it was never built in. As a result, Hamburg has today a system of open basins cut into the land, leaving solid piers projecting; it has not the English system of closed docks, with their hours of inactivity when ships and barges cannot get into the docks, or, if they are in, cannot get out.

Basins or "harbors" (Häfen) were cut into the land because the cheaper process, prevalent in America, of

THE PORT OF HAMBURG

building piers out into the water, was not practicable. Hamburg lies eighty-five miles distant from the open sea, up the river Elbe, and the river is here so narrow that the construction of projecting piers would have left insufficient width for a channel. But, for a reason which we shall consider later, most of the basins were made wide enough so that vessels could lie at the quays and discharge into freight sheds, while at the same time other ships tied up at a line of mooring posts, which bisects the basin longitudinally, and discharged into lighters and up-country barges alongside. The first of these slips or basins, the Sandtorhafen, was opened in 1866. In the seventies the Grasbrookhafen was opened.

In the meanwhile, in 1871, the German Empire had been formed, into which Hamburg and Bremen entered only on condition that they should remain outside the Customs Union, consisting essentially of the members of the Empire. Germany developed rapidly, in an industrial way, and imports and exports for it began to be of greater significance for Hamburg than the old transshipment trade. After the independence of Belgium was attained, in the thirties, Antwerp awoke to a new commercial and maritime greatness, and, by the excellence of its new port facilities and the versatility of its steamship connections, was drawing heavily on the foreign trade of West Germany. It was time for the Elbe port to prepare for the needs of modern commerce. Bismarck had long importuned Hamburg to join the Customs Union. In 1882 it consented, ostensibly unwillingly. No doubt a leading ground for its consent was the fear that the exceptional tariffs which the German railways, under the leadership of Prussia, were granting to German seaports, would be

PORT FACILITIES

withheld from one that persisted in remaining a foreign country.

However, a good bargain with the Empire was made, which retained for Hamburg many of the advantages that it had formerly enjoyed. The state of Hamburg, practically identical with the city of Hamburg, with 275,000 inhabitants, entered the Customs Union. Its harbor proper was to remain outside the Union and was to be rebuilt, isolated from the rest of the city. The Empire agreed to contribute forty million marks towards the construction of this Free Port.[a] The remaining cost—about one hundred and fifty million marks[b]—has been borne by Hamburg.

The Sandtorhafen and the Grasbrookhafen had been built into a peninsula on the right—cityward—bank of the main Elbe stream. The whole peninsula, as well as the island of Kehrwieder between it and the city, was pre-empted for the Free Port. As no one was allowed to live there, 1,000 property owners were expropriated and 24,000 people made homeless.[c] In this right-bank peninsula one more huge basin, the Baakenhafen, was constructed; 1,200 acres of marsh land were purchased on the left bank of the river and new basins excavated there. To the Free Port, opened in 1888, many additions have been made, all on the left bank, the last and greatest being

[a] Of course Hamburg sacrificed something when it came into the Customs Union. Her 900,000 inhabitants now pay the German duties —averaging perhaps 25 per cent—on imports. According to the terms of the agreement with the Empire, Hamburg must pay 1,700 men $1,000,000 a year to guard the Free Port. (Boston Society of Architects, page 24.)
[b] Wiedenfeld: Hamburg als Welthafen, page 19.
[c] Aftalion, page 505.

THE PORT OF HAMBURG

the basins on Kuhwärder, built for and leased to the Hamburg-American Line.

The Free Port consists of a large number of basins, lined by quay walls, alongside which steamers can lie and be discharged by cranes into freight sheds, amply supplied with railway connections. In the wide basins, mooring posts provide anchorage for ships handling cargo in the stream. There are warehouses directly on the waterside. Between the various left-bank basins are located shipyards and numerous exporting industries. The whole Free Port, therefore, considered by the customs department as foreign territory, includes land on either bank of the Elbe and the main river itself for a considerable distance. It is surrounded by a customs line, guarded by customs officials. On land the line is designated by high iron palings; along the river it is a floating palisade; where it crosses the river it is an imaginary line guarded at either end by the customs men. At the land and water entrances into the Free Port are provided customs booths, where goods must pay duty when they enter the Empire.

The first advantage of the Free Port is in facilitating re-exportation; indeed, the importance of the re-exportation trade is what, before all else, led to its creation. Merchandise can be brought free of duty into the Free Port, stored in its warehouses, repacked or mixed and then, as conditions of the market dictate, sent across the customs line into Germany or shipped to Scandinavia and the Baltic. In the Free Port foreign merchants can maintain sample or consignment stocks. Bonded warehouses do not offer the same opportunity for unhindered movement of merchandise within a port: everything must be done under the harassing control of customs men. In

PORT FACILITIES

Hamburg there is no need of counting and verifying pieces when a re-exportation is made. A bonded warehouse cannot offer the same facilities for various manipulations necessary to prepare goods for the consumer, such as cutting wines and mixing coffees.[a]

The privilege of manufacturing in its Free Port, which Hamburg alone of all German ports possesses, is one that has proved of less benefit than was expected. Its advantage is of course that it allows exporting and outfitting industries to get their foreign raw materials duty free. This advantage has been partly overcome by the system of draw-backs since introduced and applied to manufacturers in the Customs Union: refunding to exporting manufacturers the duty paid on foreign raw materials contained in their manufactured products. The disadvantage under which all industries in the Free Port labor is that, if they wish to sell in Germany, they have to pay on their products crossing the customs line the high duty on manufactured articles, while their inland competitor has had to pay only a low duty on the corresponding raw materials. This disadvantage has become more marked as the home has come to preponderate over the foreign market.

Excepting shipyards, the industries in the Free Port have grown incomparably slower than those elsewhere in Hamburg and are of distinct types.[b] They cater to the building, outfitting and provisioning of ships; such are

[a] As a Hamburg merchant said, it is not so simple to make Javan from Brazilian coffee, in case of need. (Wiedenfeld: Die Welthäfen, page 289.)

[b] There are about 15,000 workmen employed in the Free Port, not more than 3,000 outside the shipyards.

THE PORT OF HAMBURG

shipyards, boiler shops, machine and repair shops and biscuit factories. Or they represent industries principally interested in exporting, such as rice mills and oil mills; or industries settled in the Free Port region before the Free Port was built.[a] There has been complaint that manufacturers of inferior and "schwindelhaften" wares have sought the Free Port out, in order to be free from the severe German official regulations.[b]

Perhaps the chief advantage of the Free Port lies in the facilities it offers for the rapid, frictionless discharging of ships with dutiable goods, whether destined for re-exportation or shipment inland. As Hamburg lies eighty-five miles from the sea, precautions must be taken to prevent goods being landed on the way up. The Hamburg pilot, who must be taken aboard when the vessel enters the Elbe, is sworn in as a customs inspector. Under his guidance the vessel comes up the river at any hour of day or night and passes to her berth in the Free Port, unmolested by customs officers. There are no summary or detailed declarations of dutiable goods to be made, no customs officers to be taken aboard, with the explanations and delays attendant on their presence. Where, as in England, their official hours are limited, a ship with dutiable wares must suspend the discharge of her cargo

[a] The industries of the Free Port are located across the river from Hamburg. Provision had to be made for feeding the workmen there. This is accomplished by numerous restaurants (Kaffeehallen) under state control. As workmen may not live in the Free Port, there is maintained an elaborate ferry service between various parts of it and the city. In 1909 a tunnel was opened from Hamburg to Steinwärder, in the Free Port, where, among other establishments, the largest Hamburg shipyard, Blohm and Voss, is located.

[b] Aftalion, page 195.

Hamburg freighter at a pier, discharging.

PORT FACILITIES

during night hours. In the Hamburg Free Port, she discharges and loads day and night, if she will. When she is ready, her inspector-pilot takes her out to sea; no officer of the customs has even been aboard. It is the least conceivable hindrance of the free movement of a ship.

There is of course no occasion for vessels engaged in the German coasting trade to enter the Free Port. They discharge on quays on the right bank of the Elbe, outside the Free Port district. Moreover, as the Elbe above and the Elbe below Hamburg are within the Customs Union and as part of the main stream is within the Free Port, barges plying between upper and lower Elbe would have to be examined after coming through the Free Port, if they were obliged to use the main stream for their passage. To obviate this necessity, the right bank peninsula, already mentioned, has been cut to form a belt canal, which acts as a partial boundary between the Free Port and the city, and through which the barge traffic between upper and lower Elbe plies unhindered. However, it is in the Free Port that the main part of Hamburg's shipping is handled, and its facilities are those that interest us.

There is little essential difference between the various basins of the Free Port. A description of the newest and best, the piers of the Hamburg-American Line, will serve for all. The Hamburg-American piers were constructed by the state of Hamburg in the year 1903, at a cost of thirty-two million marks, and leased to the Hamburg-American Line at a yearly rental of 1,350,000 marks.[a] Three basins have been cut into an ancient meadow, the

[a] Richter: Führer, page 63.

THE PORT OF HAMBURG

Kuhwärder, leaving two huge, solid piers projecting. The piers are lined with concrete quay walls, with foundations so deep that the berths are dredged to a depth of 32.8 feet, the high water depth of the channel. The solid piers are so wide that each carries two rows of sheds, one on either edge, as well as railroad tracks before and behind the sheds. One side of the longer pier has a length of 3,500 feet and carries three enormous pier sheds. In front of these, spanning the water-edge railroad tracks, are numerous electric cranes—one every 100 feet— with a lifting power of two tons each. They have a "half-portal" form: a vertical leg runs on a rail on the very edge of the quay, a horizontal leg on a rail on the shed, just above the door.[a] Cars and locomotives pass beneath the cranes, which are themselves movable longitudinally. Back of the shed, flush with the rear railway tracks, is a street, which may be used to dray goods to Hamburg or to near-by piers. Hand cranes serve to lower heavy

[a] It is interesting to observe, in various German harbors, cranes representing successive stages in pier crane development. The earliest form was the stationary hand crane. Then came the stationary steam crane, disadvantageous because the vessel had to be moved to and fro to bring her hatches, one after the other, within reach of the crane. The stationary steam crane was followed by the movable steam crane—it was easier to move the crane than the ship. But the early movable steam crane took up the width of a railroad track on the pier edge, for itself. The next stage was the electric (or hydraulic) portal crane, which put the mechanism out of the way, seated on a portal that ran on rails outside the railroad tracks, which it spanned. The last stage was the half-portal crane, in which the inner leg of the portal is taken out of the way; the horizontal beam of the framework runs on the shed itself. Perhaps the highest stage of pier crane development is reached by the roof cranes at Liverpool. They run on the slanting roof of the pier shed, outside, and do not in any way interfere with the utilization of the space on the pier's edge beneath.

PORT FACILITIES

pieces from the rear platform into car or dray. For handling freight within the pier shed, the Hamburg-American Line is experimenting with electric trucks of the three-wheel type, capable of carrying 5,000 pounds each at a speed of four miles per hour. Each does the work of six men with the old hand trucks.

Before a liner arrives, the import shed where she is to discharge and the export shed where she is to load are ready for her. Freight trains and drays have been unloaded on the front and rear platforms of the export shed, river barges and harbor lighters have lain alongside the quay and had their freight hoisted and swung across the railroad tracks to the shed platform by silent electric cranes. A big river barge—they average 600 to 800 tons capacity—cannot enter the seaship basins unless it has at least fifty tons of freight to exchange with a seaship; otherwise it must send a little harbor lighter—they average about sixty tons—to the ship.[a] This prevents unnecessarily clogging up the basins with unwieldy river barges. Similarly, the import shed has been emptied for the liner by cars, lighters, drays and barges.

The liner ties up at her import berth. A swarm of pier cranes brings up goods from the hold and swings them across to the shed platform, whence they are trucked inside. There they are counted, sorted and arranged for shipment direct inland by barge or rail, or for sending by lighter or dray to warehouse or railroad station (less-than-carload lots). If freight needs no sorting, it may

[a] Recently there is talk of increasing to 100 tons this minimum that a barge must have for a seaship in order to be allowed to approach it. Surely such regulation will be necessary for the harbor of New York, when the Barge Canal is finished and the Barge Canal Terminal is constructed in New York.

THE PORT OF HAMBURG

be dropped by the crane into a freight car standing underneath. Similarly, goods destined per barge for points on the upper Elbe, on through bills of lading, may be dropped by the ship's tackle overside directly into the barge.

In the Berlin Museum für Meereskunde (Marine Museum) is a model of one end of a basin of the Hamburg-American Line, on Kuhwärder. The "Patricia" and the "Blücher," in the New York service, lie at the Auguste Victoria quay, before shed 73, which is about 1,300 feet long and 200 feet deep. Half of the building is serving as import, half as export shed. The "Blücher" has nearly completed her cargo and the shed behind her is almost empty. On her water side a great floating crane has brought and is lowering into a hatch huge steel beams for bridge construction work, too heavy for the pier cranes or the ship's derricks to handle. The ship's derrick is raising from a lighter, which has hastened up, an express consignment that nearly missed the boat. A fresh water boat has come alongside and is charging the ship's fresh water tanks. On the land side a line of coal cars stands on the quay's edge. Boards have been laid from the cars to chutes into the ship's bunkers, and over these boards men are walking with baskets, coaling the vessel.

The "Patricia" is discharging. Here cranes are in full activity, swinging cargo from the ship's hold across to the shed platform, there to be trucked inside. A part of the roof of the model has been removed and through the opening one sees merchandise in the shed being sorted. Already they are loading into freight cars from the rear platform of the shed, and into drays which will take the goods to Hamburg. On the water side of the vessel, the

PORT FACILITIES

ship's tackle is lowering freight into lighters. Among them lies a grain discharger, opposite a portion of the ship filled with grain; it lies between the ship and an 800-ton Elbe barge. The discharger lets its long proboscis down into the hold, sucks out the grain, cleans and weighs it and slides it into the capacious hatches of the river craft.[a] A remarkable expedition of discharge is attained by the use of all this freight-handling machinery, particularly the pier cranes. A ship like the "Patricia," which, besides a long passenger list, carries a cargo of 10,000 tons, is unloaded in about forty hours and loaded in thirty to forty more.[b] This is at the rate of 250 tons of cargo per hour and is the regular rate of discharge at the Kuhwärder piers. On December 9, 1910, the "Saxonia" of the Cunard Line created a new speed record for handling cargo in Boston. She loaded 4,500 tons in twenty-six hours, a speed of 175 tons per hour.[c]

At the Reiher quay, the repair berth across the end of the Kuhwärder basin in the model, lies the pleasure yacht "Prinzessin Victoria Luise," since lost. That she is being repaired is indicated by the scaffoldings of scrapers and painters on her hull and funnels. The big twenty-ton hammer-shaped crane on the quay is lowering into her a boiler; next the crane lie heavy pieces of machinery, awaiting their turn. At the vessel's prow a lighter is testing the anchor chain, link by link.

[a] For discharging loose grain in Hamburg, there are ten floating pneumatic elevators, which are capable of discharging 700 to 800 tons per elevator per day, out of one hatch. Thus four of them working on a ship discharge 3,000 tons per day.
[b] Stahlberg, page 30.
[c] See Boston papers of December 10.

THE PORT OF HAMBURG

It has been observed that the mooring posts of early days have been retained in the new basins. The basin in the model is wide enough to accommodate a line of vessels at either quay and a line on either side of the double row of mooring posts that runs down the center of the basin. In the model, the "Rhenania" and "Abyssinia" are tied up at the mooring posts. They are loading with their own tackle from a horde of lighters and barges that surround them, bringing cargo from the railways and from up-river. Perhaps there is no room for them at the quay. Perhaps they are not in line service, but are engaged in taking a casual cargo of bulk goods of low specific value, such as potash or raw sugar. In the case of state piers, wharfage dues are purposely put so high that they discourage the use of piers by tramp steamers with bulk cargo. In the case of the "Rhenania" and "Abyssinia," the Hamburg-American Line apparently thinks that their cargo is not of sufficient value and does not require such expedition in loading as to justify them in taking up room at the piers. In general, the liners, carrying package freight, which demands expedition, discharge and load at the piers.[a] The vessels in casual or tramp service, whose cargoes, as a rule, consist of bulk goods, handle their freight in midstream. It is of course possible to discharge a cargo of package freight at the piers and go to the mooring posts to take on a bulk

[a] An addition to the pier dues of a steamer, after she has been at a state pier for five days, causes all the larger steamers which dock there to go to midstream to load. They cannot discharge and load in five days, and the additional dues are too heavy for them to stand. The lines which lease piers have both discharging and loading of their liners done at the piers.

PORT FACILITIES

cargo. In any case, the midstream mooring posts mean a doubling of the port's capacity.[a] Nearly half the tonnage of vessels entering Hamburg discharge in midstream, as is indicated by the following table:

VESSELS USING PIERS AND MOORING POSTS TO DISCHARGE.
HAMBURG, 1907.[b]

	Vessels	Reg. Tons	Average Tonnage per Ship
State Piers,	5,023	3,903,000	777
Leased Piers,	736	2,389,000	3,246
All Piers,	5,759	6,292,000	1,093
Discharged at Mooring Posts,	10,714	5,748,400	536

Ten thousand seven hundred and fourteen vessels of 5,748,000 register tons did not use the piers to discharge. The difference between the average size of ships discharging at the state and at the leased piers will be noted. Lines to England, Scandinavia, etc., are berthed at the state piers; their steamers run from 500 to 1,000 tons. The great oversea companies with large steamers—such as the Hamburg-American Line, the German East African Line and the Woermann Line—lease their piers. In general, state ownership prevails in the older, right-bank basins, whose shallower depth suffices to accommodate the smaller European liners. Leased piers predominate in the newer, deeper basins on the left bank, and in the right-bank Baakenhafen. At the state piers, vessels are

[a] When the berths are crowded, even steamers with general cargo must often discharge at the mooring posts. The operation is slower there than at the quay, owing to the necessity of assorting packages on the ship's deck, according to marks and numbers, and owing to the necessity of constantly shifting lighters.

[b] Statistik der Kaiverwaltung. Hamburg, 1907.

accommodated in turn, preference being given, however, to steamers over sailing ships, and to steamers of regular lines over irregular visitors.

English coal (four and one half million tons in 1907) is discharged in the new coal "Hafen," below the Kuhwärder basins. Discharge is by tackle and basket. There is a separate Petroleumhafen, whose entrance can be closed by an iron pontoon, which prevents burning petroleum from flowing out into the main harbor; the pontoon is kept so closed at night. The shores of this basin are lined with tanks, into which oil is pumped from the tank steamers of the Deutsch-Amerikanische Petroleum Aktien-Gesellschaft, or as it is abbreviated, "Dapag," the German daughter of Standard Oil. One of its steamers, the "Niagara," carries 10,000 tons of oil.

There are strangely few grain elevator buildings at the water's edge, scooping their grain out of ships; but, as we shall see, the grain trade and grain storage have moved inland. At the coal quay, coal tips grasp a car and tip it until its contents flow into a ship. The tip is used principally in loading coal for export, and for coaling tramps, which are little disturbed by the dirt occasioned by having their bunker coal dropped into them from a height of several yards and which do not mind the loss of time in going to and from the coal quay. Liners, which usually carry passengers and cleanly freight, cannot stand the dirt or lose the time. While lying at their berths, loading, they are coaled from lighters on the water side or coal cars on the land side. There has been recently constructed in the Free Port a great potash elevator, somewhat on the principle of the American grain elevator building. Its bucket-chain brings the potash up from

PORT FACILITIES

Elbe barges. It is stored and later slid into ships for export.

The entire Hamburg harbor consists of eighteen such basins or "harbors" as those of the Hamburg-American Line, already described. The depth of these basins varies from five and one half meters (eighteen feet) in the Sandtorhafen, the oldest, to ten meters (32.8 feet) in the new Kuhwärder basins. There are also numerous basins of the harbor, yet to be described, devoted to its river barges, and having depths of four to six feet and more. The water surface of the harbor has grown from 61 acres in 1854 to 1,576 acres in 1909.[a] This acreage was used as follows:

WATER AREA OF HAMBURG HARBOR.

Basins for seaships,	723 acres
Basins for river barges,	375 acres
Canals and branches with seaship depth,	36 acres
Canals and branches with barge depth,	103 acres
Main stream and entrance to basins,	338 acres
Total,	1,576 acres

The entire length of water front is forty-one miles. Of this total, twenty-one and one half miles border on water with a depth for seaships, nineteen and one half miles on water with a depth for river barges. Of this total water front of forty-one miles, thirteen and three fourths miles have been supplied with perpendicular quay

[a] These statistics are from: Hamburg als Schiffahrts- und Industrieplatz, I. Beiblatt, Hamburg, 1910. In addition to the water space detailed above, there are fifty-four acres of protected water area in Cuxhaven. The total area of the Free Port, including the land therein, is 2,508 acres. (Barge Canal, I., 412.)

THE PORT OF HAMBURG

walls with deep foundations, which allow ship or barge to come alongside, within reach of the cranes.

The combined length of the harbor pier sheds is eight and one half miles, their total floor space five million square feet. There are 808 cranes in the harbor, with a combined lifting power of two million tons. The largest cranes lift 150, 75 and 50 tons, respectively. The length of railroad tracks in the harbor is 138 miles—by way of comparison, the distance from Hamburg to Berlin is 189 miles.

There are four lines of transportation engaged in collecting and distributing the freight carried by the seaship: dray, lighter, barge and railroad car. The following table illustrates the part played by each of these vehicles in disposing of goods discharged at the piers in Hamburg:

REMOVAL OF GOODS DISCHARGED AT HAMBURG PIERS, 1907.[a]

By	Per Cent
Rail,	18
Dray,	20
Lighter,	45

That leaves 17 per cent removed by barges, whose cargo, however, is procured mainly from vessels in midstream.

Drays are used primarily to carry goods between the piers and the city of Hamburg, such as goods for local consumption. The local consumption of a city of 900,000 inhabitants is a very considerable item. Horse and wagon also take merchandise from warehouse to the freight-receiving station of the railroad. In winter,

[a] Report of the Metropolitan Improvements Commission, page 185. Boston, 1910.

PORT FACILITIES

when ice in the harbor makes lightering dangerous, drays are called on to transport goods between piers, as in the transshipment trade, or between pier shed and Free Port warehouse.

There are in the harbor of Hamburg 500 covered and 2,500 uncovered lighters, 3,000 in all,[a] with a carrying capacity of 20 to 250 tons each. As already set forth, they were formerly carried by the tide from ships anchored in midstream, which had given them their goods, to warehouses lying on the old city canals, and vice versa. They were guided and helped along by their steersmen, supplied with long poles with hooks in the end. The pole could be put into the river and pushed on, or hooked to an anchored vessel or anything else solid, and the lighter thus pulled forward. This method of procedure was borrowed from London, where it may still be observed in full bloom. In London, the custom of towing lighters is just coming in, principally for tows of coal upstream. But Hamburg has no potent guild of Watermen and Lightermen who have prevented lightering from emerging from its mediæval form of organization. The Hamburg lighters are towed about. The stream is not cluttered up with struggling, swearing lightermen, drifting helplessly with the tide, in the way of each other and of everyone else. There are numerous Hamburg lighterage companies, and this business is also a branch of the large inland transportation concerns, notably the United Elbe Navigation Company.

Lighters are engaged in transferring goods between pier shed and warehouse—practically all warehouses, in the Free Port and in the city, are on the water's edge;

[a] Barge Canal Terminal Commission, I.. 411.

THE PORT OF HAMBURG

between pier shed and pier shed or ship and ship, in the transshipment trade; between pier shed or midstream ship and the waterside freight-assemblage station of the railroad, soon to be described. Finally, the lighters mediate between barges and seaships which have less than fifty tons of cargo to exchange with each other, in which case the barge is not allowed to approach the ship.

Elbe barges handle over half the freight that Hamburg sends inland or receives from there. They are primarily engaged in carrying bulk goods—coal, grain, potash, raw sugar—which they interchange usually with vessels moored in midstream. However, they handle a very large amount of freight in interchange with the ships at the quays. If their liner is at her berth, they can lie alongside and interchange cargo over the ship's side. If she is not at her berth yet, or has left, they lie at the quay and have their freight discharged or loaded by pier cranes. The nature of the barge terminal facilities at Hamburg is more fully described in a later chapter.

Finally, the railroads. The harbor tracks are owned by Hamburg but operated by the Prussian state railways, which form Hamburg's connection with Germany, as part of the Prussian system. Delivery to the port of Hamburg means delivery to any pier; there is a terminal charge of one mark (23.8 cents) per ton, reduced on certain low grade goods discharged direct between car and ship at quays especially fitted up for this form of transfer—supplied with cranes but no pier sheds. However, this traffic is inconsiderable;[a] by far the larger part of

[a] It varies from 100,000 to 150,000 tons of goods per year in all, and consists primarily of Chili saltpetre received, and coal shipped. (Stahlberg, page 26.)

Pier cranes discharging cargo at Hamburg.

PORT FACILITIES

the freight arriving by sea, and discharged at the quays, is sorted at the pier sheds before shipment inland. If rail shipments are in carload lots, they are sent direct from the shed platform. If they are in less-than-carload lots, they are put on lighter or dray—on lighter, if they come from one of the new, distant piers—and sent to the assemblage freight-receiving station of the railroad.

There is a large difference in the railroad tariffs on package freight in carload and less-than-carload lots, respectively. So in Germany an important rôle is played by the forwarders, who assemble L. C. L. shipments and send them off by the carload. Part of the saving in freight rate, which they effect, is of course allowed to the shipper. These forwarders are allowed space in the assemblage freight-receiving station, located near the Free Port warehouses, on the water's edge, so that it is accessible to dray or lighter. Each forwarder sends so often per week a carload or more of this freight to Berlin, one to Dresden, to Munich, etc. Each steamship company and warehousing concern has its forwarder. The counterpart of the railroad assemblage freight-receiving station, where forwarders send in carload lots combinations of L. C. L. shipments, is a freight-delivery station, similarly situated. Here the forwarders receive in carloads assembled shipments from their agents from different points in Germany and near-by foreign countries. The separate shipments are sorted out and drayed or lightered to their destinations all over the harbor.[a]

[a] Transit cars from Austria, Switzerland, etc., with goods destined for export via Hamburg, are sealed when they cross the German border. Their seals are simply removed when they enter the Free Port. If a car has assembled L. C. L. shipments, for export, it goes

THE PORT OF HAMBURG

The important warehousing business of the Free Port is in the hands of a privately operated, partly state owned and state controlled concern, the Free Port Warehousing Company (Freihafen-Lagerhausgesellschaft).[a] Before 1885 Hamburg's warehouses lay on the various city canals. When the city proper became part of the Customs Union, the old warehouses became simply bonded warehouses and there was need for the erection of the freer type of warehouse in the Free Port. Hamburg decided that these should be under state control. The North German Bank of Hamburg was authorized to establish a storage concern, the Free Port Warehousing Company, under terms agreed upon by the financial department of the city. The first buildings were erected in the Free Port upon public land, the Kehrwieder island. They still stand, a handsome row of red sandstone structures, facing the city and looking more like a row of university building than like warehouses.

The company was empowered to issue warrants, transferable to order, on goods stored on the property. The stock capital was fixed at nine million marks[b] ($2,142,-000); tariffs for storage, handling and manipulating goods were fixed in the contract between city and company. Hamburg put 322,930 square feet of land at the disposal of the company and undertook to build the necessary quay walls and slips, in return for a share in the profits. From 1889 to 1905 Hamburg received 3½

to the assemblage delivery station of the railroad and has its packages sorted out and delivered, all without customs oversight or interference. No bonded warehouse can offer such facilities.

[a] There is a fairly detailed report on Free Port warehouses in the Barge Canal Commission's Report, I., 412-416.

[b] 1889-1905, dividends were 5 per cent; since then, 5½ per cent.

PORT FACILITIES

per cent—since 1905, 5½ per cent—on a capital of fifteen million marks, representing the value of the land, quays and slips which it furnished. In addition, a portion of the net profits each year is set aside to create a fund for acquisition of the company's stock by the state, which will eventually be full owner.

The few original warehouses did not long suffice. Sixteen million marks of bonds have been issued to build new ones. By January 1, 1911, 1,161,170 square feet of ground were covered by warehouse buildings, affording 5,417,725 square feet of storage space. Three fifths of this space is leased by the company to particular firms; the remainder is operated by the company in its capacity as a storage concern.

The warehouses are built in double rows, between which a lighter canal runs. Lighters lie alongside and have their merchandise hoisted direct to the floor of the warehouse to which it is destined. Re-exported goods are similarly transferred by lighter from warehouse to seaship. No duty is paid until goods cross the customs line into Germany. Warehoused goods destined inland by barge are lightered to the freight shed of an Elbe navigation company. If destined inland by rail in carload lots, they are shipped direct from the warehouse door;[a] less-than-carload lots are drayed to the near-by assemblage freight-receiving station of the railroad. If destined for Hamburg for local consumption, goods are drayed across the bridge over the Zoll Canal, which

[a] The older warehouses have no direct rail connection. This is comparatively unimportant; most inland shipments of warehoused goods are in less-than-carload lots and must be drayed to the railroad freight station.

THE PORT OF HAMBURG

separates the Kehrwieder from the business heart of the city. The largest and finest warehouse, the Kaiserspeicher, has not only rail connection but also a depth of water alongside such that ships can come and be discharged direct into the warehouse: there is no lightering necessary.

The storage business of the company developed slowly. The great importers were loath to leave off their custom of employing their own little warehousing concerns on the canals, who had become expert in performing the necessary manipulations for them and in making their shipments. But the greater freedom of the Free Port warehouses as compared with the—now—bonded warehouses in town; the offer of the company to lease to the merchants space in the new warehouses, where their old agents could still carry out orders for them; the convenience which the company's warrants afforded as security for loans, etc.; the greater safety and hence lower insurance premiums on goods in the new buildings;[a] all this finally attracted the warehousing business to the Free Port.

GROWTH OF STORAGE BUSINESS OF THE FREE PORT WAREHOUSING COMPANY.[b]

	I. Packages in Storage	II. Bags of Coffee (included in I.)	III. Receipts for year. Storage Department
End of 1889,	345,624	147,137	619,980 Marks
End of 1898,	825,116	603,012	1,115,150 Marks
End of 1908,	1,927,617	1,761,847	2,401,610 Marks

[a] The company takes out a general insurance policy on its warehouses and goods stored therein. In consideration of average monthly premiums, it places these policies at the disposal of the merchants. (Barge Canal, I., 414.)

[b] Barge Canal, I., 414. At the end of January, 1910, the coffee on hand amounted to 2,236,306 bags.

PORT FACILITIES

Particularly worthy of attention is the emigrant village erected by the Hamburg-American Line on land furnished free by the Hamburg government. We shall later see the significance of the Hamburg emigrant trade in furnishing the ships bound for America with a human return freight.

The emigrant village lies on the left bank of the Elbe, opposite Hamburg and completely segregated from the city. This was done in order to protect Hamburg from contagious diseases. The need of such protection became particularly apparent after the cholera epidemic of 1892 and after the majority of the emigrants had come to consist no longer of Germans but of Russians and Austro-Hungarians. A secondary aim of the village is to protect its sojourners from extortion at the hands of the Hamburg merchants.

The village was built in 1900-01 at the cost of three million marks. It consists of about twenty-five buildings, accommodating five thousand persons and is designed to receive only the emigrants arriving from countries where the standard of health is low. At a nominal charge these emigrants are here sheltered from the time of their arrival until the departure of their ship.[a] They have already been examined, on entering Germany. Those diseased and those whose physical, moral or financial status promises their rejection at the hands of the American immigration officials, are rejected before they cross the German border, lest they later fall a burden to the German steamship company which would have to bring them back from America. In solid trains those who pass

[a] Twenty-five per cent of the emigrants pay nothing. The others pay from fifty pfennigs to one mark per day, each.

THE PORT OF HAMBURG

the border inspection are brought to the emigrant village and there re-examined, to eliminate any cases which have been overlooked or which have developed on the journey.

The emigrants are received in a huge inspection building, where the few suspicious cases are weeded out and sent to the "observation pavilion" across the road. Most of the arrivals are at once passed to the pavilion where they are to live, though some must first bathe and have their clothes disinfected. A feature of the emigrant village is the simple hotel, where for a slightly higher price the better class of emigrants can have hotel accommodations.[a] Each pavilion consists of a dormitory, a large living room, baths, etc. Nationalities are carefully kept separate. There is one large dining hall, with a section set aside for the Jews. The Jews also have their separate kitchen whose methods are supervised by an appointee of the chief rabbi of Hamburg. The village contains a synagogue, a Catholic and a Protestant church. German emigrants are, in general, not received in the village; they must stop in the licensed boarding houses in Hamburg.

The careful surveillance exercised over these boarding houses, and over merchants and others who are tempted to swindle emigrants, the Imperial inspectors who scrupulously inspect the emigrant ships and, above all, the excellent accommodations which the Hamburg-American Line offers foreigners in its village—all these factors conspire to make Hamburg a very popular point of departure for the European emigrant. The degree of

[a] Similar to this is the creation of a third-class passage on the boats of the Hamburg-American Line, with accommodations between those of the second class and the steerage.

PORT FACILITIES

that popularity is expressed in the high percentage which the Hamburg-American Line gets of the proceeds of the transatlantic emigrant pool, in whose hands the emigrant trade lies. Many an emigrant departs with regret from the village, just as many of them prefer a slow boat to a fast one because the slow boat prolongs a steerage passage where they enjoy a scale of living such as they never knew before.

Finally, there are the harbor works at Cuxhaven, at the mouth of the Elbe, eighty-five miles below Hamburg. Cuxhaven was acquired by Hamburg in the fourteenth century.[a] The situation commanded the entrance to the Elbe, on which Hamburg's life depended. Moreover, Cuxhaven was an excellent harbor of refuge, to which vessels could run in from the stormy North Sea. At the end of the nineteenth century it began to look as if it would be impossible to dredge the Elbe deep enough for modern liners to continue to come up to Hamburg; already they were having to lighter a considerable por-

[a] In 1246, half of the island of Neuwerk had been given to Hamburg by the Bishop of Bremen, on condition of the erection of a lighthouse thereon. Later, acquisition was made of the rest of the island, on which a Hamburg deputy, the Ratsherr, lived and watched for pirates and shipwrecks. Opposite the island, the disagreeable lords of Ritzebüttel lived in a castle, supporting themselves and their retainers by the pursuit of piracy. In 1393 the Hamburgers made an expedition against the robber lords and captured the castle. To forestall possible reprisals, Ritzebüttel and the neighboring land were bought from the robbers. From the castle of Ritzebüttel, piracy was now as zealously suppressed as it was formerly practiced. A small refuge harbor was built and pilots were here taken aboard. When the course of the Elbe shifted and left Ritzebüttel high and dry, "Kuckshafen" was built near by, on the new bank of the stream. There was for centuries difficulty in keeping it from shifting away from Cuxhaven. (Buchheister, pages 173 seq.)

THE PORT OF HAMBURG

tion of their cargoes before proceeding up the river. There were already two basins at Cuxhaven; in 1895 a third was constructed. The Hamburg-American Line, owner of the largest ships plying to Hamburg, asked the state to construct and lease to it a quay of this basin with pier shed, cranes, etc., and with a large railway station and customs office. The intention seemed to be to transfer to Cuxhaven the freight and passenger terminus of the big liners engaged in the New York trade.

These constructions were carried out and leased to the Hamburg-American Line in 1902 for twenty-five years at a yearly rental of 111,000 marks. While the construction work was going on, the steamship company acquired land in Cuxhaven, and began to erect houses for the captains and other ship's officers, foremen, clerks and workmen who would in the future be attached to the new terminal.

There is a ten-foot difference between high and low tide at Cuxhaven, and soon after the new basin, with its great depth, was opened, it was found that it exhibited a strong tendency to fill up with mud. Moreover, it was found to be dangerous to enter or leave this open basin except at high or low water because of the violence of the tidal flow at other times. A storm in November, 1903, tore the "Deutschland" from her berth at the quay and inflicted severe damage on her; which demonstrated that there was here too little protection for loading and unloading at a pier shed. At the same time it was becoming apparent that the Elbe could, after all, be dredged for ships of the deepest draught. So the Hamburg-American Line abandoned the use of all parts of the harbor excepting the railway station. Incoming

PORT FACILITIES

passengers are put off the liners at Cuxhaven and get inland by train several hours earlier than if they had steamed up the river. Mails are of course put off with the passengers. The same time is saved by outgoing passengers and mails, which are not put aboard until the liner reaches Cuxhaven. In good weather the ship comes alongside the river quay, near the railroad station, but outside the basin; in bad weather passengers and mails are transferred by lighter.

The cost of constructing the harbor, including the harbor works at Cuxhaven (nine and one half million marks), but not including dredging the channel nor erecting the Free Port warehouses, has amounted, to date, to about four hundred million marks.[a] In return for this expenditure Hamburg does not receive much in the way of direct dues. According to the budget for 1906, as presented to the Senate, there was expected the following income from the harbor:[b]

INCOME OF HAMBURG HARBOR, 1906.

Total, ... 4,861,000 Marks

---

[a] Die Seeinteressen des deutschen Reiches, Teil IV., and Die Entwicklung der deutschen Seeinteressen, Teil VI. In 1909, forty-five million marks more were voted, to carry out extensions. Below the Kuhwärder harbor four new basins will be built: two for seaships, each 300 meters (984 feet) wide; one for seaships, 210 meters (689 feet) wide. Separated from these will be a new Petroleumhafen, 140 meters (460 feet) wide. The present chain of barge basins, behind the left-bank seaship basins and connecting with them, will be continued to serve the new extensions. Between this chain of barge basins and one of the new seaship basins, there will be room for a new shipbuilding yard. (Barge Canal, I., 415.)

[b] Richter, page 28.

THE PORT OF HAMBURG

PRINCIPAL ITEMS.

Dues from state piers,	2,757,000 Marks
Rentals from leased piers,	1,903,000 Marks

The two principal items are thus seen to be: dues collected for the use of state piers, and rentals from leased piers. The third item would be harbormaster's dues, collected from vessels using the mooring posts, and levied according to their draught. These dues are purposely low, to encourage vessels with bulk cargoes to discharge in midstream instead of at the crowded piers.

The quay dues on cargo discharged at state piers are one mark per metric ton. Ships discharging part of their cargo on the pier and part overside, need pay cargo dues only on the tonnage actually going over the quay. These dues entitle the ship to the use of the pier, pier shed, cranes and cranemen, and storage in the pier shed for two days. Stevedores to discharge the cargo must be found by the ships. In addition to these cargo dues, the vessel must pay seventeen and one half pfennigs per net cubic meter (11.8 cents per net register ton) for the first five days at the berth; three and one half pfennigs per net cubic meter (2.36 cents per net register ton) for each succeeding day, excluding holidays and Sundays.

Ships discharging at the mooring posts pay only the harbormaster's dues of five marks ($1.19) when of not over two meters (6.56 feet) draught, and five marks for each additional meter. Thus the largest ship that could get up the Elbe would have to pay only forty-five marks dues for discharging in midstream. It is an insignificant item; the cheapness of discharge at the mooring posts is a great boon to vessels with bulk cargo. Other dues are

PORT FACILITIES

for pilotage (one pilot to Brunsbüttel, one thence to Hamburg), perhaps sanitary dues, etc.

The following are the major items of expense accruing to a vessel of 3,100 tons gross, 2,000 tons net, loaded full with a cargo of dead weight, and departing in ballast:

CHARGES ACCRUING TO A VESSEL DISCHARGING IN HAMBURG.[a]

Total, ...	1,769 Marks=$421.18
PRINCIPAL ITEMS.	
Admiralty pilotage, . . .	244 Marks
Pilotage from Brunsbüttel, .	81 Marks
Tonnage dues, ...	679 Marks
Harbor towage, in and out, .	257 Marks
Pilotage outwards, . . .	123 Marks
"To attendance to ship's business,"	215 Marks

Stevedore's charges for unloading the cargo are not included in the above total, though harbormaster's dues are. The reputable stevedores in Hamburg have combined and have identical charges. For instance, the charge for unloading "Bombay cargoes" in midstream is 70 pfennigs per ton. Assuming that our ship had 3,600 tons of such cargo aboard, it would cost 2,520 marks ($510) to discharge her. The total cost to her of entering and discharging her cargo in midstream would be $421.18+$510=$931.18.

The stevedore's charge for unloading Bombay cargoes at the pier is 45 pfennigs per ton.[b] If the vessel discharges there, she pays 3,600 marks ($857.14) of cargo dues, and 1,091.20 marks ($236) dues levied according to the ship's net cubic measurement. The total

[a] Barge Canal, I., 400.
[b] Less than for discharge in midstream; cranes are supplied with the pier.

THE PORT OF HAMBURG

cost to her of entering and discharging her cargo at the pier is 421.18^a+857.14+$236=$1,514.32$. It is apparent why bulk cargoes, of low specific value, are handled in midstream.

In the 1902 Report of the Royal Commission on the Port of London, Sir Alfred Jones, chairman of the board of directors of Elder, Dempster & Company, gave the following comparison of the costs accruing in various European ports to a freight steamer of 5,146 net register tons, with a cargo of 5,000 tons of grain, 3,000 tons of package freight and 2,000 tons of wood:

DUES IN VARIOUS EUROPEAN PORTS.

	Liverpool	Hamburg	Rotterdam	Antwerp	Bremerhaven	London
Harbor dues,	£404	£411	£136	£228	£347	£368
Unloading costs,	522	442	310	404	385	592
	£926	£853	£446	£632	£732	£960

But such a comparison has a doubtful value. The relative dearness of a harbor is determined not only by these items but also by consulate dues, the cost of lightering the shipment in the harbor from ship to river barge or of switching it over the harbor belt railway to its point of departure inland, etc. To compare the ports on this scale is not possible. Jones says: "The conditions affecting the various ports are so different that a useful comparison is almost impossible." Sir Thomas Sutherland, chairman of the Peninsular and Oriental, made little of the height of the dues and said: "Rapid discharge is the most vital of all questions." Jones compressed a volume of criticism of London in the sentence: "While London

[a] Minus the small harbormaster's dues contained in this total.

In the emigrant village at Hamburg. Catholic church.

PORT FACILITIES

costs about the same as Liverpool or Avonmouth, the despatch is about five times as bad in the case of a large vessel in London."[a] Despatch is life to the ocean liners. The port of London, with the seven and one half million inhabitants of the great city as its assured dependents, may be able to afford not to keep pace with its rivals; certainly no other port can. In this regard, despatch, Hamburg enjoys an enviable reputation. It has been seen that the "Patricia," with 10,000 tons cargo, is unloaded in forty hours, loaded in forty more, a feat which probably no other port can equal.

Hamburg, then, has supplied itself with the most modern harbor facilities: provision for the rapid transfer of freight between the ocean, river and rail carriers. Direct contact between the carriers is secured; there is no unprofitable and dilatory juggling of the freight necessary in order to get it from one to the other. The Free Port lets the Hamburg merchants store their goods duty-free and offers them complete freedom of manipulation and the desired option of re-exporting them or of sending them inland, as the market dictates. A still more important advantage of the Free Port today, is that it allows ships in the foreign trade to discharge with the utmost freedom and expedition, without customs officials causing them the least harassment or delay. Industries in the Free Port labor under a positive disadvantage, excepting those directly catering to ships and those more interested in foreign than home markets.

Proper provision has been made for the various special functions which the port has to perform: warehouses, the emigrant village, the port of call at Cuxhaven, the petro-

[a] Report of the Commission, page 26.

THE PORT OF HAMBURG

leum harbor, etc. The entire port is operated on a financial plan less calculated to make it a profitable investment than to bring prosperity to the city. As compared with other great European ports, Hamburg's dues are not high; in respect of that vital need, despatch, none stands higher.

CHAPTER IV

Hamburg's Oversea Lines

PART of the equipment of a port are its oversea steamship connections. They are the port's messengers to the most distant lands; they seek out the purveyors and customers of its importers and exporters and inland manufacturers. In general, the great desiderata in ocean transportation are cheapness, safety, speed and regularity. Under modern conditions cheapness and safety—the lack of the latter would appear in high insurance rates—are approximately equal for the fleets that serve first-class ports. These ports fight the battle for supremacy with the weapons of speed and regularity in their oversea services. Frequent and speedy oversea connection between Hamburg and a foreign country often determines whether a German export consignment is sent via Hamburg, or via Bremen or Antwerp.[a] If the foreign purchaser can get the goods more quickly from England, he may employ neither the German manufacturer nor the German steamship line.

Steamship connection between Hamburg and foreign ports falls into two periods, which are divided by the year 1871, the date of the foundation of the German Empire. At this date Hamburg had one oversea line:

[a] De Rousiers, page 218, repeats the remark of an Elberfeld exporter that he did not always export via Antwerp and Rotterdam. He exported to Manila and Mexico via Hamburg, for he had a sailing once a week from Hamburg, once a month from Antwerp and still less frequently from Rotterdam.

THE PORT OF HAMBURG

the line of the Hamburg-American Company to New York. For part of its trade with other ports it was dependent on casual (tramp) service, offered only when a shipload of goods, usually bulk goods, presented itself. For regular connection, such as the importer of valuable wares or the manufacturer or his exporter must have, Hamburg was dependent on England, principally London. This transshipment traffic accounted for a large portion of its trade with England, and we have seen that in 1869 arrivals from England amounted to 60 per cent of the tonnage of ships entering Hamburg.

Such casual and indirect connections are the rule in all trade relations which are not enough developed to support direct, regular lines. The fact that half the steam tonnage of the world is still in tramp steamers testifies to the demand that still exists for casual service. The large transshipment trade of Hamburg and London to Scandinavia and the Baltic serves those ports that cannot yet exchange regularly full shiploads with the particular countries to which they transship via London and Hamburg. It was from such a state of dependency as this that Hamburg had to free itself.

The significance to a port of direct oversea connection with all parts of the world is perhaps better understood in Europe than in America. European countries were for centuries dependent on London; in the last thirty years there has been a race for freedom among them. When Hamburg exported and imported via London, the English middleman invariably took toll on all that passed through his hands. The German manufacturer paid more than the English to get his goods to market: he paid the costs of shipping his goods to England and

HAMBURG'S OVERSEA LINES

transshipping them there. If the English liner had a full cargo, it was the German goods that waited for the next boat. Moreover, sales are often preceded by a considerable correspondence. There were frequent complaints regarding the delaying of German mails sent by English boats. But the speedier steamers of regular lines are necessary for more than the mails. German export industries have so much invested in them that money cannot lie idly tied up in their products, waiting for a tramp to get a full cargo. Many goods are exposed to serious deterioration in a long voyage; in the case of others the duration of transportation is an important factor in determining the selling price. Many orders for manufactured goods stipulate immediate delivery. Exporting industries and regular steamship lines are indissolubly bound up together.

The advantages of regular line service can be partially attained by foreign companies coming in to serve a port. But foreign companies come in only when there is already enough indirect trade between the terminals of the new line to assure its success. They offer no service to create that trade; they do not speculate. They are ready to withdraw the moment that the financial success of the undertaking becomes doubtful. This is not true of a native line, financed and believed in by the merchants and capitalists of the seaport. Even if the foreign line becomes fixed, it never attains that degree of personal acquaintanceship, trust and consideration which obtains between merchants and a line domiciled in their own port. Finally, the educative and advertising value of steamship lines is lost if a country's products are not carried abroad under its own flag. A great steamship is a splendid evidence of the industrial power and ability of the nation it represents.

In Chapter I. we sketched the change in Germany's economic life: her development into an "industrial" state, highly dependent on foreign trade. German agriculture demands fertilizers, yet even so it cannot support the population of the Empire. Foreign lands must send foodstuffs and the raw materials of industry; to pay for them Germany exports manufactures.

Under these new conditions it was ridiculous for Germany or Hamburg to be dependent on casual tramp service or indirect steamship "line" connection with foreign ports. As the trade relations, sketched above, developed, direct German lines were created to meet them, nor did these latter always wait until the volume of trade promised a certain profit from the establishment of a line. For instance, in 1871, the Hamburg-American Line established a service to the West Indies, which remained a losing investment until 1879.[a] For years the same fate met its line to North Brazil.

Early in the nineteenth century the merchant carrier still prevailed. Great Hamburg merchant houses owned one or more sailers; the builder of the vessel and its captain often had a share in her. It was considered part of the dignity of these houses to have such ships representing them; it spread abroad their name and prestige just as a merchant marine today carries the flag and prestige of its country—as nothing else does—out into the world. But the work of ocean transportation became too big for a branch of a mercantile establishment to handle.

[a] Prof. K. Thiess: Die Hamburg-Amerika Linie, page 18. Berlin, 1906.

HAMBURG'S OVERSEA LINES

It is now the business of separate transportation companies. Elder, Dempster and Company in Liverpool and the Woermann Line in Hamburg are two survivals of the ancient order of things: both bear the name of great mercantile houses. Yet it is significant that in both cases the mercantile has been completely severed from the transportation department.[a] The enormous capital necessary to finance a modern steamship line has clamored for the change from private or semi-private ownership to the stock company. All the larger Hamburg lines excepting the Woermann Line (to West Africa) are stock companies.

A marked characteristic of the Hamburg and Bremen lines is the large proportion of their stock held by local merchants, capitalists and bankers. It is estimated that over half the stock of the Hamburg-American Line is held in Hamburg, as over half the stock of the North German Lloyd is held in Bremen. This explains the establishment of lines that promise no immediate profit. It explains the extent to which outside companies have been compelled to abandon Hamburg. The Stettin Lloyd ran from 1869 to 1876 between Stettin and New York, calling at Hamburg. It was driven out of Hamburg and failed. The Hamburg-American Line established a service Stettin-New York in its stead. In 1902 the Booth Line of Liverpool, which had maintained a service between Hamburg and Antwerp and the Amazon, found it unprofitable to continue this line after the Hamburg-American Line had entered the same field. The close community of interests between banks, mer-

[a] Wiedenfeld: Die nordwesteuropäischen Welthäfen, page 209.

THE PORT OF HAMBURG

chants and steamship lines has been a source of strength to all three.

Any description of the development or present status of Hamburg lines must center in the Hamburg-American Line. Not only is it the largest steamship company in any country, but it comprises half the ocean shipping of Hamburg, affords a far larger proportion of her connections with oversea and is actively interested in all the other larger lines excepting the German Australian Steamship Company. The Hamburg-American Line was established in 1847; its official name is the Hamburg-Amerikanische Paketfahrt Aktien-Gesellschaft. The initial letters of the official name spell the word Hapag and it is as Hapag that the line is popularly known.[a] It is a convenient abbreviation to employ, just as the Lloyd is a convenient abbreviation for the great Bremen company, the North German Lloyd.

The Hapag was founded in 1847 to prevent a further concentration in Bremen of the American mail service, as well as imports from America of cotton and tobacco and exports thither of German emigrants. Bremen had had a packet sailing line to New York since 1826. The Hapag began business with three copper-bottomed sailing ships of together 1,600 register tons, and with a capital of 460,000 marks. The "Deutschland," which left on her first voyage to New York on October 15, 1848, was of 717 tons register and had room for 20 cabin and 200 steerage passengers.

[a] In view of the luxury that prevails on the company's New York boats and the prices which passengers must pay for it, it is suggested that the initial letters mean, "Haben alle Passagiere auch Geld?" (Are all passengers well supplied with money?)

HAMBURG'S OVERSEA LINES

Until well into the nineties the transportation of steerage passengers played the chief rôle in the New York business of the Hapag and the Lloyd and its profits enabled those companies to build up their fleets.[a] In the early days of the Hapag there was little return freight to America to exchange for the bulk products we sent Germany. But Germany did send us men. The average number of steerage passengers brought us by the two German companies in the years 1860-1900 was as follows:[b]

YEARLY AVERAGE OF STEERAGE PASSENGERS.

From Hamburg		From Bremen	
1861-70	34,466	1862-71	45,213
1881-90	90,889	1882-91	97,909
1891-1900	60,041	1892-1900	61,379

German emigration to the United States, in those years since 1870 when it exceeded 100,000, was as follows:[a]

GERMAN EMIGRATION TO THE UNITED STATES.

1870	110,000	1884	179,000
1873	149,000	1885	124,000
1881	210,000	1887	106,000
1882	250,000	1888	109,000
1883	194,000	1891	113,000
	1892	119,000	

In 1895 the German emigration dropped to 32,000 and has never since reached 50,000. But, as the German

[a] Thiess: Die Hamburg, page 26.
[b] Fitger: Die wirtschaftliche und technische Entwicklung der Seeschiffahrt, 1902, page 19.

THE PORT OF HAMBURG

emigration died down, the Hapag and Lloyd concluded a treaty with the other continental transatlantic lines, whereby the German lines were to have the transportation of East European emigrants to the United States. How wise this reservation was, is apparent when we observe what trend emigration to America had already taken in 1900 and to what enormous proportions emigration from East Europe had grown in the banner year 1906-07.[a]

EMIGRATION TO UNITED STATES FROM CONTINENT OF EUROPE.

Year	Germany	Russia	Austria	Italy	Total
1870	118,000	1,000	44,000	2,000	387,000
1882	250,000	21,000	29,000	32,000	788,000
1892	119,000	61,000	76,000	61,000	379,000
1900	18,000	90,000	114,000	100,000	448,000
1907	37,000	258,000	338,000	285,000	1,285,000

The part the German lines play in the transportation of these hordes is very large. In 1907 they landed in New York from their German and Italian services the following number of emigrants:

EMIGRANTS LANDED BY GERMAN COMPANIES IN NEW YORK.[b]

Hapag, 150,633   Lloyd, 160,574

Their nearest rival was the Cunard Line, which in 1906 landed in New York from its British and its Mediterranean services 107,790 steerage passengers.

[a] Report of the Commissioner of Immigration in the 1907 Report of the Department of Commerce and Labor.

[b] 1907 Report of the United States Commissioner of Navigation, pages 146-7. The German companies were hard hit when the American panic of 1907 set in and ruined their emigration business for 1908. Emigration via Hamburg dropped to 78,808 in 1908; in Bremen it dropped to 74,626. (Nauticus, 1909, page 298.) In 1910, emigration

HAMBURG'S OVERSEA LINES

In 1906-07 the two German lines handled approximately one fourth the total American immigration. Assuming an average fare of $35.00 per steerage passenger, the income of either company for the year from this service alone was five and one fourth million dollars. Germany now exports largely to the United States, but her exports are primarily manufactured articles of high specific value and small bulk. The ship room occupied by American goods cannot be filled for the return voyage with German goods. Ships are so equipped that certain decks may serve as cargo space for the east-bound trip and in Bremen and Hamburg be transformed into steerage quarters for the west-bound. The fact that emigrants can be counted on as return freight has had a great influence on the inducements that the German companies could offer in freight rates from the United States to the continent. It accounts, in large measure, for instance, for the concentration in Bremen of the European trade in American cotton and tobacco.

The Hapag's three sailing vessels of 1848 maintained a monthly service with New York, averaging forty-

had about regained its normal status. In the year ending December 31, 1910, there were landed in New York, by the three leading companies in the passenger and emigrant business, the following number of persons:

Cabin and Steerage Passengers Landed in New York, 1910.

Line	Cabin Passengers	Steerage Passengers
Hapag,	40,021	114,023
Lloyd,	49,307	111,517
Cunard,	37,878	93,312

(Figures of the Landing Agent of the United States Immigration Service, published in New York papers, January 11, 1911.)

one days on the ocean west-bound, twenty-nine days east-bound. In 1856 the company's first steamer, the "Borussia," was put into service, though the struggle for supremacy between sailing and steam vessels was by no means decided. Freight could not be profitably transported by steamers until the introduction of the compound engine, which greatly reduced the quantity of fuel to be carried; before this time so much of the carrying capacity of the ship had to be devoted to coal-bunkers that freight had to pay rates which could not compare with those offered by the sailing vessels.[a] But, as we have seen, the North Atlantic trade was preëminently a passenger trade. The saving in time which the steamers afforded—though slight at first—their greater steadiness and safety, conspired to give them the preference in the Hapag's fleet. By 1867 the last of the sailers was sold. The Lloyd, founded in 1857, never possessed a sailing vessel.

Business for the Hapag was excellent in the decade 1860-70. Especially was this true for the years 1865-66-67. The American Civil War was over and commerce was renewed with the Union, which needed supplies to repair the devastation that had been wrought. The American merchant marine had been destroyed and foreign

[a] In 1866 Adolph Wagner wrote an article on Ocean Transportation in Rentsch's Handwörterbuch der Volkswirtschaftslehre. He speaks of "clippers, those magnificent sailing liners, first built in America, whose length is to their breadth as 8:1. This relation was formerly 3:1 or 4:1 in sailing ships. Hamburg clippers like the 'Donau' have made the trip from New York to Cuxhaven in 18 days, while good steamers do not get under $13\frac{1}{2}$ to 14 days and ordinary sailing vessels take 5 or 6 weeks." Wagner thought that the steamers would eventually have the transportation of persons and package freight, while the sailing vessels retained bulk freight.

HAMBURG'S OVERSEA LINES

carriers came into its heritage. In the three years 1865-66-67 the Hapag distributed 20, 20 and 16 per cent in dividends, respectively.

The year 1871, when the Empire was formed, saw the Hamburg lines multiply. The Hapag instituted a service to the West Indies; in the same year the Hamburg South American Steamship Company was founded to ply between Hamburg, Brazil and the la Plata. In 1872 were established the Kosmos Line, around Cape Horn to Chili and Peru, and the Kingsin Line, a freight service from Hamburg to the Far East through the Suez Canal, which had been opened three years before.

The panic of 1873 set in and held up further advances. In 1873 the Hapag fell into a rate war with the newly founded Adler Line, established to partake in the profits of the transatlantic trade. It was a war that lasted three years. The Hapag, which had paid 12, 16 and 20 per cent dividends in the years 1871-73, paid no more dividends until 1878. The contest ended in 1875, when the Hapag unwillingly purchased the Adler steamers. In order to do this, the company's capital was increased from sixteen and one half to twenty-two and one half million marks; it had to be scaled down to fifteen million in 1877. The Hapag was left with a fleet out of all proportion to its needs, into which it could not grow for years to come. Yet the cloud had a silver lining. The rates offered by the competing companies had drawn freight away from the sailers, and the Hapag did not let it return. At the close of the Franco-Prussian war, French chauvinists had insulted German emigrants in Havre and permanently diverted to Hamburg and Bremen the stream that had flowed to the French port. More-

THE PORT OF HAMBURG

over, Hamburg was learning from Bremen how to handle and attract emigrants. Early in the century Bremen had recognized the future of the emigrant trade. There were enacted in Bremen severe regulations against ill-housing, underfeeding, swindling or otherwise maltreating emigrants,—matters to which Hamburg was too long indifferent. This policy gave Bremen on the entire continent a reputation that still endures and is worth to her thousands of emigrants yearly.

Hardly had the Hapag recovered from its first rate war when, in 1883, a second broke out. The new Edward Carr Line and the old Hamburg shipowner, Robert Sloman, who had instituted as early as 1849 an emigrant sailing line to New York, united to form the Union Steamship Company, which fought the Hapag's line to New York. The struggle was terminated in 1886, when the Hapag and the Union companies combined their schedules. In 1888 the Carr steamers were purchased; the 1886 agreement with Sloman ran on until 1907, when his steamers also were bought. This second rate war gave the Hapag its present brilliant manager, Albert Ballin, who was taken over from the Carr Line; and it taught the company the value of timely compromise in ocean warfare. Since then the Hapag has been the prime mover in many of the combinations that go to make up the complicated system of agreements, pools, defensive and offensive alliances and fusions that prevail in ocean shipping today.

From now on the Hapag had a clear field. Foreign trade grew: Hamburg's hinterland demanded more and more grain, meat and other foodstuffs, fertilizers and fodder and the raw materials of industry; it exported

The "Deutschland," the Hapag's only express steamer, paid for with the proceeds of the sale to Spain of obsolete Hapag liners, at the time of the Spanish-American war. She is being built over into a pleasure cruiser, the "Victoria Luise."

HAMBURG'S OVERSEA LINES

more and more potash, sugar and manufactured goods as Germany became established in the markets of the world. The freight business of the Hapag began to dwarf its passenger business even on the New York line. In 1891 the Hamburg Hansa Steamship Company was bought up and its lines to Montreal, Boston and Philadelphia turned over to the Hapag flag, which was already serving Baltimore. In 1898 a freight line was established to China and the Far East and the Hamburg Kingsin Line (1871) of the same destination was purchased.

In the mean time other lines had not been idle. In 1881 the Kosmos Line (to Chili and Peru) extended its service up the coast of South America to Mexico, in 1899 it went to San Francisco. In the following year the Hapag made the Kosmos Line let the Hapag share its extremely profitable service to West America under an agreement whereby either company supplies a certain proportion of the total number of steamers despatched per year. Profit and loss are distributed according to the tonnage of steamers which each company supplies. It is a common form of agreement in Germany and one by which the all-powerful Hapag has come to have a share in nearly every other profitable steamship company in Hamburg. The Kosmos also has a line from Genoa to West America and in the course of its career has bought up the rival Hamburg Pacific Steamship Company.

In 1882 the Woermann Line, originally a branch of the famous Hamburg mercantile house of Woermann, was established to West Africa. Since 1907 the Hapag has also shared this service. The Lloyd and the Hamburg-Bremen Africa Line had formed a similar partnership. In 1907 they entered into a "community of interest"

THE PORT OF HAMBURG

(Interessengemeinschaft) with the Hapag and Woermann. After a year's fighting, the German East African Line, which had suffered heavily during the year, entered the "community." Thus all German lines to Africa are united.

When in 1910 the Hapag and Woermann decided to establish a direct connection between New York and West Africa, they announced the following program:[a] "The steamship Carl Woermann will begin in December a new and direct service between this port and the west coast of Africa, under the joint auspices of the Hamburg-American and Woermann lines. Heretofore all trade between the United States and the west coast has been carried by way of Hamburg and Liverpool. The Canary Islands will be the first port of the service. The ships will call at a hundred or more places on the west coast. Very little money will be used. The ships will exchange oil, tobacco, flour, machinery, cotton goods and food products—practically every ship will be a department store afloat—for mahogany, palm oil, rubber, ivory, cacao and copper. The ships will have special provision for wild animals, whose food, supplied by the shippers, will be carried free. It will cost a shipper $250 to bring an elephant to New York; $200 for a giraffe; $100 for a lion, tiger or leopard, and $25 for an ostrich." To the uninitiate, the fare for giraffes seems particularly reasonable.

What motive may influence the Hapag in a move of this sort is indicated by the rise and fall of the line from New York to the Levant (through to Constantinople and

[a] New York papers, November, 1910.

HAMBURG'S OVERSEA LINES

Odessa), maintained by the Hapag and the German Levant Line in common, 1901-04. The considerable traffic from New York to the Levant, transshipped at Hamburg, was threatened by the prospect in 1901 that a direct Russian or Italian line would be established. So the Hapag and the German Levant Line instituted the direct connection themselves and continued it for three years. In 1904, when conditions had changed and there was no longer fear of Russian or Italian competition, the line was withdrawn; New York goods for the Levant were again brought to Hamburg and transshipped there to the steamers of the German Levant Line.[a]

In 1888 the Hamburg-Australian Steamship Company came into life. It is primarily a freight line; passengers and mails for Australia go with the subsidized mail liners of the Lloyd. The Hapag has no connection with the Hamburg-Australian; there was long, apparently, an understanding between the Hapag and the Lloyd that Australia should be left to the Lloyd, in return for which the latter kept her hands off Africa.

In 1890 the German East African Line was established with a subsidy of 900,000 marks yearly, given to assure regular connection between Germany and her colony of German East Africa. The Hapag and Woermann are now financially interested in the line. At first it ran through the Suez Canal and down the east coast as far as Delagoa Bay. Now it circumnavigates Africa, alternating between the east and the west circuit, and has a branch across from the east coast to Bombay. Its subsidy amounts to 1,350,000 marks per year and a considerable part of its service is performed by the Hapag.

[a] K. Thiess: Die Hamburg-Amerika Linie, page 34.

THE PORT OF HAMBURG

The Hamburg South American Steamship Company had prospered continually since its birth in 1871. In 1900 it was engaged in a rate war with the De Freitas Line, also of Hamburg, which with a fleet of fourteen steamers maintained a service from Hamburg to the Amazon. In 1900 the Hapag bought the De Freitas steamers and signed a treaty with the Hamburg South American, which still holds. By its terms the South American services of the allied companies are specified as four: to North Brazil, Central Brazil, South Brazil and the Argentine. For these lines the Hapag furnishes one third of the steamers, the Hamburg South American two thirds. The inland and foreign agencies of the Hapag book passengers for both companies.

Such is the history of the foundation of Hamburg lines other than the Hapag. Since 1900 the chronicle has mainly to do with the Hapag. In 1897, at its fiftieth jubilee, it could announce that it was the largest steamship company in the world. Since then it has strengthened its position. It owns a network of feeder lines on the Chinese coast and, in common with the Lloyd, a river line on the Yang Tse Kiang. The Hapag trades to all Chinese ports, its freighters reach to Nagasaki and Vladivostok. The lines to the Far East have received particular attention since the 1901 tour of inspection of the Hapag manager, Ballin, to these regions. The Chinese freighters, returning, call at Calcutta and share the large transports of the Bremen Hansa Line from Calcutta to Europe. The last few years have seen the establishment of a new Hapag service to Arabia, Persia and the Soudan.

HAMBURG'S OVERSEA LINES

The West Indies were since 1871 a chosen field of the Hapag. It was the first service established after the original line to New York; in 1901 the Lloyd had also entered these waters. In this year the services of the two to Mexico and Cuba came into collision. The Lloyd withdrew from Mexico, the Hapag from Cuba, excepting that her Mexican liners call at Havana to land or take on mails and passengers. In 1901 the Hapag bought out the seven steamers of the English Atlas Line plying between New York and the West Indies. In the treaty of 1902 with the Morgan Ship Trust—of which more later—the Hapag stipulated that no member of the Trust should enter the New York-West Indies service. The Leyland Line, with a heavy trade between England and the West Indies, was meant in particular. In 1902 there was trouble with the Cameron Line, New York to Hayti; it was paid to withdraw and leave its officials in the service of the Hapag. But in 1906 a competitor entered the race who was not so easily disposed of: the Royal Mail Steam Packet Company, which had plied from London (later Southampton) to the West Indies since 1839. Competition between it and the Hapag (which ended in 1908 [a]) raised the service from New York to the West Indies to a high standard: steamers of 10,000 tons and over. Lastly, the Hapag has a treaty with the United Fruit Company, by which the Hapag shares in the transportation of bananas from Costa Rica.

There is a treaty between the Hapag and the Kansas City, Mexico and Orient Railway, according to which the Hapag has the privilege of establishing—as soon as the

[a] Nauticus, 1909, pages 3, 4.

THE PORT OF HAMBURG

railway is completed—a service from Port Stilwell, its Mexican terminal, to Asia and Australia. This would complete the steamship company's ring around the globe.[a] Beginning with 1904 four Hapag steamers were chartered by the Portland and Asiatic Steamship Company.[b] In 1898 the Hapag and an English company instituted a monthly service from New York to the Far East through the Suez Canal.[c]

The service to the Argentine from Genoa is an important one. It was instituted in 1896. The Italian government legislated in favor of Italian companies; so the Hapag founded an Italian company under the name "Italia" and turned over to it the line from Italy to the la Plata. This company improved its steamers and established a coasting line that served the entire peninsula of Italy and the neighboring French coast; in 1905 the Hapag sold its share therein to the Navigazione Generale Italiana. Immediately thereafter, legal discrimination in favor of native lines ceased and the Hapag renewed its former line. It now plays an important part in the heavy trade between Italy and the Argentine. Besides the freight trade, there is a large movement of Italian laborers to the Argentine before each year's harvest; they return when the harvest is over. The line Italy-

[a] K. Thiess: Die Hamburg-Amerika Linie, page 53.
[b] Himer, page 100.
[c] Of course the Hapag does not establish lines with its eyes shut. Typical for the Hapag, as for German procedure, was the trip of investigation which Ballin, the general manager, made in China in 1901, and which resulted in a considerable extension of the Hapag lines thither. The Lloyd has had similar investigations of economic conditions and possibilities made in Australia and in the South Sea Islands, the latter in 1907.

HAMBURG'S OVERSEA LINES

New York, originally a passenger line for winter travelers, has seen its center of gravity shift to the freight and the emigrant trade. In the winter there is a weekly Egyptian Express run from Berlin to Naples to connect with a Hapag express steamer, which plies to Alexandria. Moreover, numerous cruises to the Mediterranean from Hamburg and New York are instituted. The Lloyd having already allied itself with Thomas Cook and the Cook hotels in the Orient, the Hapag was obliged to purchase from Carl Stangen a tourist agency of its own and build its own hotels in Palestine. It is part owner of the steamers of the Hamburg and Anglo-American Nile Company.[a]

Another view of the magnitude of the Hapag's undertakings is afforded when we consider its subsidiary operations in South America. Brazil has coasting laws like our own, reserving the coasting traffic to vessels carrying the Brazilian flag. So the Hapag, the Hamburg South American *et al* organized in 1905 a Brazilian coasting company under the name Companhia de Navagaçao Cruzeiro do Sul, to act as feeder of their South American lines.[b] In the same year they established an Argentine coasting line from Buenos Aires to the southern end of Patagonia. In 1905 the Hapag, in consequence of a treaty with the Venezuelan government, took over the re-organization of the Venezuelan coasting trade.[c] In 1906 the Hapag took shares in the Santa

[a] It is even said that the Hapag and the Lloyd have secured for themselves coal fields in Rhineland-Westphalia, to assure their future supply of bunker coal, at reasonable rates. (Haarmann, page 90.)
[b] Himer, page 100.
[c] Thiess: Die Hamburg, page 53.

THE PORT OF HAMBURG

Katharina Railway Company, formed to run from the coast to the German settlements in South Brazil.[a]

The 300,000 German settlers in South Brazil are not to be deserted. Now and then a strange light pierces the darkness of German official circles and illumines the outside world. On November 28, 1899, at the time the present German naval program was being campaigned for in Germany, one high in the councils of state spoke in public as follows:[b]

"We must desire that at any price a German land with a German population of twenty to thirty millions shall arise in South Brazil in the course of the next century. It is indifferent whether it becomes an independent state, or whether it comes in close connection with our Empire. Without trade connections, constantly made secure by warships,—without the possibility of Germany showing her power there decisively, this development is threatened with failure."

In addition to all these lines to distant countries, there is an abundance of connection between Hamburg and the neighboring lands of Europe. The Hapag maintains a heavy service with sea-going lighters to Copenhagen and ports of the Baltic, to Bremen, to Emden, up the Dortmund-Ems Canal to Dortmund, and up the Rhine as far as Strassburg. Largely by their use of sea-going lighters to collect and distribute goods, Hamburg and Bremen have shattered the maritime hopes of Emden

---

[a] Himer, page 44.

[b] Geheimrat Professor Dr. von Schmoller, Exzellenz, member of the Prussian House of Lords, in a speech at the Philharmonic Hall, Berlin. The speech is reprinted in the collection, "Handels- und Machtpolitik," by Schmoller, Wagner and Sering, Stuttgart, 1909, page 35.

HAMBURG'S OVERSEA LINES

and German Baltic ports.[a] The Hapag and the Rheinschiffahrts-A. G. vorm. Fendel issue a circular that reads as follows: "The Hamburg-American Line, in conjunction with the Rheinschiffahrts-A. G. vorm. Fendel of Mannheim, maintains a direct and regular barge service from Strassburg, Karlsruhe and Mannheim-Ludwigshafen to Hamburg. Departures every sixth day, with direct, through bills of lading, via Hamburg, to transatlantic places." Besides these sea-going lighters, which also go up the Dortmund-Ems Canal as far as Dortmund (125 miles), there are Rhine-sea steamers, which ply between Hamburg and points on the German Rhine as far up as Cologne (210 miles from the sea). Hamburg sent up to the German Rhine, or received from it, over 300,000 tons of freight in 1907.[b] Robert Sloman goes to Italy, the Neptune Steamship Company (of Bremen) serves the entire European coast, the Hamburg-Petersburg Line runs to Russia. In these near services, foreign and outside German companies play an important part. The Svenska Lloyd, the Oldenburg-Portuguese Steamship Company (to Spain and Portugal) and the Leith, Hull and Hamburg Steam Packet Company are such. Nor are there foreign lines lacking in the distant routes. The

[a] *Growth of Sea Lighter Tonnage in German Ports.*

	1896		1900		1909	
	Vessels	Gross Tonnage	Vessels	Gross Tonnage	Vessels	Gross Tonnage
All Germany,	136	31,761	178	53,468	313	106,358
Bremen,	80	21,474	102	31,386	143	52,264
Hamburg,	51	9,721	58	19,933	135	47,245

(Cords, 34-35 and Statistisches Jahrbuch, 1909.)

[b] Jahresbericht der Zentralkommission für die Rheinschiffahrt, 1907.

THE PORT OF HAMBURG

Stoomvaart Maatschappij Nederland and the Rotterdamsche Lloyd call regularly at Hamburg on their way to the East Indies, the Union Castle Mail Steamship Company on its way to South Africa, the North German Lloyd en route to Australia.

Such are, in outline form, the line connections of Hamburg with the outside world. Hamburg owned on January 1, 1910, a merchant marine of 1,570,635 net register tons (673 steamers with 1,304,315 net tons, 512 sailing vessels with 266,050 net tons). The corresponding total gross register tonnage was 2,312,775.[a] The larger companies in Hamburg stood as follows on the same date:[b]

LARGER HAMBURG STEAMSHIP COMPANIES, 1910.

	Capital	Steamers	Gross Register Tons
Hamburg-American Line,	125 mill. marks	178	872,820
Hamburg South American Steamship Company,	15 mill. marks	38	187,692
German "Kosmos" Steamship Company,	14 mill. marks	37	183,429

[a] Hamburg als Schiffahrts- und Industrieplatz. I. Blatt. January, 1910. The average value of the stocks of the principal Hamburg companies on the years 1905-07 was as follows:

Average Value of Stock of Hamburg Steamship Companies.

	1905	1906	1907
Hamburg-American,	156	161	134
Hamburg South American,	157	168	141
German Kosmos,	163	184	176
German Australian,	128	128	115
German Levant,	89	84	69
German East African,	86	85	71

(Hamburgs Handel und Schiffahrt, 1907.)

[b] Hamburg als Schiffahrts- und Industrieplatz, January, 1910.

HAMBURG'S OVERSEA LINES

	Capital	Steamers	Gross Register Tons
German Australian Steamship Company,	16 mill. marks	33	144,358
Woermann Line,	Private	33	80,670
German American Petroleum Company,	9 mill. marks	21	84,467
German East African Line,	10 mill. marks	18	67,715
German Levant Line,	6 mill. marks	26	62,166
SAILERS.			
Steamship Company of 1896,	2 mill. marks	21	40,478
F. Laeisz,	Private	16	39,485

Hamburg has 54 per cent of the total gross register tonnage (steam) of the German merchant marine.[a] When we consider the large part played by the fast-going line steamers in the Hamburg tonnage, it is apparent that its portion of the carrying power of the German merchant marine is still larger.

Unfortunately, there is nothing later than Wiedenfeld's compilation to show the relative frequency, in North European and English ports, of regular steamship service to foreign parts. The compilation was made in 1901 and since then all the ports have increased their services, but they have done this in unison and it is unlikely that their relative positions have changed. In

[a] 3,889,000 tons (Lloyd's Register for 1909-10). A characteristic of the German merchant marine is its concentration in the hands of few companies. Eighty per cent of it is in the hands of nine companies; the Hapag alone represents 30 per cent. The White Star Line, the largest British company, with its 461,000 tons, represents but an insignificant percentage of the British merchant marine of nineteen million tons. In England the tramp still predominates, in Germany the liner. (De Rousiers, pages 119-120.)

THE PORT OF HAMBURG

1901 there were the following number of regular steamship sailings per week:[a]

REGULAR SAILINGS PER WEEK FROM EUROPEAN AND ENGLISH PORTS.

TO	From						
	London	Liverpool	Hamburg	Bremen	Amsterdam	Rotterdam	Antwerp	Havre
New York,	1-2	3-4	3	2-3		1	2	1¼
Other U. S. Atlantic coast ports,	3	6-7	1-2	1	½	2-3	2	
Canada,	1-2	2-3	¼				½-1	
West Indian ports,	½	2-3	2	1	½		2-3	3
Brazil,	¼	2	3	½			2-3	2
Argentine,	½	3	1-2	1	¼		2-3	¾
American west coast,		1	1				1	¾
West Africa (Cape Verde-Walfisch Bay),	1-2	2-3	2	¼		1		¼
South Africa (Walfisch Bay-Delagoa Bay),	2	1	1	¼	¼		1	
East Africa (Delagoa Bay-Aden),			½	¼	¼	¼	¼	½
India,	3	3-4		1			2	
East Indies,	½	1	¼	¼	1-2	½	½	
East Asia (North of Hong-kong),	2	½	1	½-¾		¼	2	¼
Australia,	3	½	¾	¾			1½	

It is observed that Hamburg, Liverpool and Antwerp have the most complete connections, the most versatile schedules. The distinction between them is, that Hamburg and Liverpool are served by their own steamship companies, Antwerp is a port of call for all the ships that sail the seven seas.[b] One sees, on the Scheldt, liners of

[a] Wiendenfeld, Die Welthäfen, page 205.
[b] Of course Antwerp's proximity made it a natural port of call for lines trading to England. Hamburg, 300 miles to the eastward, was more or less driven to establish its own lines.

HAMBURG'S OVERSEA LINES

the Hapag, the Lloyd, the Union Castle Line, the Kosmos Line, The Peninsular and Oriental, Lamport and Holt, the Nippon Yusen Kaisha and a hundred others. Antwerp has no great line of its own; the Red Star Line is owned by the (American) International Mercantile Marine Company and the Compagnie Internationale du Congo is financed by British capital.

The following is the status of the world's larger steamship companies; it shows the predominance of the German lines:[a]

WORLD'S LARGEST STEAMSHIP LINES.

Time	Line	Steamers	Gross Register Tons
Jan., 1910.	Hamburg-American Line,	168	934,000
Jan., 1910.	North German Lloyd,	134	679,000
Jan., 1910.	White Star Line,	30	461,000
Jan., 1910.	British India Steam Navigation Company,	111	452,000
Jan., 1910.	Ellerman Lines (including Bucknalls),	113	450,000
Jan., 1910.	Peninsular and Oriental,	58	412,000
Jan., 1910.	Alfred Holt and Company,	65	381,000
Jan., 1910.	Elder, Dempster and Company,	113	345,000
Dec., 1909.	Nippon Yusen Kaisha,	79	307,000
Dec., 1908.	Messageries Maritimes,	66	296,000
Jan., 1910.	Union Castle Line	41	295,000
Jan., 1910.	Navigazione Generale Italiana,	108	290,000
Jan., 1910.	German Hansa Steamship Company (Bremen),	56	272,000
Dec., 1908.	Compagnie Generale Transatlantique, ("French Line")	70	269,000
Jan., 1910.	Furness Line,	96	268,000
Jan., 1910.	Leyland and Company,	42	294,000
Jan., 1910.	Cunard Line,	21	234,000

[a] Hamburgs als Schiffahrts- und Industrieplatz. II. Blatt., May, 1910. The world's steamship companies having the largest number

THE PORT OF HAMBURG

The tonnage of the Hapag in this last differs from that in the list of German lines alone; here the ships in course of construction are included. By way of comparison, the steam tonnage of the American merchant marine engaged in the foreign trade in 1910 was 782,517 tons gross register.[a] That is smaller than the tonnage of the Hapag and not much larger than the Lloyd. In December, 1910, the Hapag was building fourteen steamers of 110,000 tons in all, certainly far more than we were constructing for use in the foreign trade.

It is still true that Hapag's lines to North America are its most important. Ballin, at the general meeting of the stockholders in 1908, told them that ships to a total value of one hundred million marks were in this service; that is half the book value of the Hapag fleet

of screw steamers of over 2,000 gross register tons and twelve or more knots speed, are as follows:

World's Large and Fast Steamships, 1910, of 2,000 or More Gross Register Tons and Twelve or More Knots Speed.

Line	Number of Vessels
Lloyd,	71
Hapag,	67
Messageries,	61
Peninsular,	52
Navagazione,	48
Lloyd Austriaco,	43
Nippon Yusen Kaisha,	39
Charente Steamship Company (British),	39

(1910 Report of the United States Commissioner of Navigation, pages 115 seq.)

These steamers represent one fourth of the world's sea-going tonnage, one third of its efficiency, and carry 85 per cent of the world's foreign passenger trade.

[a] 1910 Report of the Commissioner of Navigation, page 212.

HAMBURG'S OVERSEA LINES

and it includes its largest and finest steamers. Of its 127 ocean steamers in 1904, 36 were employed in the North American, 56 in the Caribbean and South American, 35 in the Asiatic routes.[a]

Out of the Hapag's single line to New York, in 1870, has grown a network of lines that serve every continent but Australia. Particular attention has been paid to perfecting the South American service, with prompt recognition of the part South America is to play in international trade. Brazilian coffee was seen supplanting East Indian; Argentine began to take the place of the United States as a purveyor of foodstuffs and fodder to Europe; with the increase in intensive farming Chili saltpetre was imported in ever increasing quantities. The purchasing power of these people rose and made them a splendid market for European manufactures.

There are several distinct advantages in this diversity of service. At certain seasons vessels can be withdrawn from one service and put in another more profitable one; for instance, the popular winter Italian-New York service can be increased at the cost of the cold northern route. Old steamers can be relegated to the less important routes while those on the more important lines can be constantly renewed.[b] The company is independent of

[a] Thiess: Die Hamburg-Amerika Linie, page 51.

[b] The average age of the Hapag ships in 1904 was seven and one half years. In that year, out of twenty-eight million marks profits only nine million were distributed as dividends. Nineteen million marks went to "write off" ships (i.e., was spent for new ones) or into the reserve fund, which in 1909 amounted to 36,800,000 marks. In 1905 the average age of the steamers of the Hapag was only 6.5 years, of the Kosmos Line 6.4, of the Lloyd 6.89 years. (Haarmann, page 78.)

THE PORT OF HAMBURG

the state of prosperity of one country or the profits of one route.[a] If a short grain crop should produce hard times in the United States, it would probably mean good times in Argentine: European importers would have to pay higher prices for the Argentine grain. The Lloyd has pursued a similar policy of decentralizing its service.

In 1901, in spite of the heavy decrease in the North Atlantic trade, in consequence of the failure of our cotton and corn crops, both the German companies distributed 6 per cent dividends. In 1904, after the rate war with the Cunard Line, which ran through the summer and autumn and destroyed all profits in the North Atlantic passenger and emigrant trade, the Hapag paid 9 per cent dividends; the Lloyd, which is more dependent on this passenger and emigrant trade, distributed only 2 per cent. In 1907, in addition to the American crisis, which became a world crisis, another rate war with the Cunard broke out. The Lloyd distributed $4\frac{1}{2}$, the Hapag 6 per cent. The crisis continued through 1908. In this year the Lloyd not only paid no dividends, it had to sacrifice its reserve fund of seventeen and one half million marks. The Hapag paid no dividends but pulled through and in 1909 was again distributing 6 per cent.

The Hapag has a happy faculty of doing the right thing at the right time. We shall see, and have seen, many examples of this in considering the agreements behind which it has entrenched itself. In 1898 it sold two of its half obsolete liners, the "Normannia" and the

[a] In the years 1902-04, when America's grain harvests were disappointing, the Hapag was in a position to remove steamers from the American service and strengthen its new lines, without building new ships. (Wiedenfeld: Hamburg als Welthafen, page 40.)

View looking into the main basin of the Hamburg-American Line, at Kuhwärder. Notice the three rows of steamers, the central row at the mooring posts.

HAMBURG'S OVERSEA LINES

"Columbia," to Spain as auxiliary cruisers. The latter had no opportunity to use them and sold the "Columbia" back to the company at a great loss. With the proceeds of the sale of these two vessels the "Deutschland" was built; she held the "blue ribbon" of the Atlantic for years. In 1904 the company sold a number of its old liners and freighters to Russia for use as auxiliary cruisers in the Japanese War. Again there was money for a comprehensive renovation of the Hapag fleet and, in addition, 11 per cent dividends were distributed in 1905. The year 1904 also brought the company considerable profits for coaling and provisioning the Russian fleet; in the following year there were heavy exports to the Far East to repair the havoc of war. The Woermann Line is said to have made its fortune from chartering its vessels to the English government as transports at the time of the Boer War.

These two wars also drew a large number of tramp steamers out of the merchant service, which made rates for those remaining in that service so profitable that an enormous overproduction in tramps resulted. The overproduction was not fully realized when the tramps first came back into the merchant service; the universal prosperity of the years 1905-07 still kept them all fairly busy. But when the crisis of 1907 broke, evil days set in. Hundreds had to be laid up, hundreds of others still sail the seas, looking for a cargo at any price.

The position of Hamburg steamship lines is strengthened by numerous agreements and alliances, among themselves and with foreign lines. These understandings run all the way from mere rate agreements to service-sharing and profit-sharing arrangements—as is frequent in the case of German lines serving common territory—or fusion. The rate agreements with foreign lines are known as conferences. Practically all the outbound freight from Europe is carried at conference rates. Rates on freight inbound to Europe, mostly bulk freight—raw materials and foodstuffs—are not fixed; the liners must be left free to meet the competition of the tramps for these goods.

The Atlantic Conference, between the North European companies on the one hand, and the English companies on the other, fixes cabin passenger fares on Atlantic liners, according to the age, etc., of the different liners. There is a territorial division of the outbound European emigrant trade, between the English and the continental lines. To the English lines is left the Scandinavian and English emigrant trade; they institute no new services on the continent, whose emigrant trade belongs to the continental companies.[a] These latter are united in the North Atlantic Steamship Lines Association and they have pooled the emigrant trade from the continent. For instance, the Cunard Line, which in 1904 established a line Fiume-New York and after a heavy rate war was admitted to the Association, gets 6 per cent of the proceeds of the pool.[b]

Probably the most important of the Hamburg agreements is that between the Hapag and the Lloyd, on the one hand, and the International Mercantile Marine Company, on the other. In 1901-02 the Leyland Line, the

[a] The White Star Line, which already had a line Italy-New York when the Conference was formed, and the Cunard Line, which broke away and established a service Fiume-Italy-New York in 1904, are exceptions to this rule.

[b] Thiess: Die Hamburg, page 45. On this subject see particularly Thiess, Himer, Passow and Wiedenfeld.

HAMBURG'S OVERSEA LINES

White Star Line, the Dominion Line, the Red Star Line, the American Line and the Atlantic Transport Line were united by J. P. Morgan into the International Mercantile Marine Company. His plan was to get control of, or agreements with, all lines between eastern North America, England and northwestern Europe. At the same time that this fusion was occurring, a twenty-year treaty was entered into by the International and the Hapag and Lloyd. The Germans and the International agreed not to enter each other's territory and to defend each other in rate wars. To a certain degree, the dividends of the German companies are guaranteed. One fourth of the amount paid by the German companies in dividends above 6 per cent is to go to the International; the latter pays the Germans one fourth of the sum lacking to give their (the German) stockholders 6 per cent on their stock.[a] How advantageous this has been for the German companies in the lean years, 1907-09, is apparent from the following statement of their dividends:

DIVIDENDS OF THE HAPAG AND LLOYD.

	1907	1908	1909
Hapag,	6%	0	6%
Lloyd,	4½%	0	0

The Holland-American Line was bought up and 51 per cent of its stock is held by the International and the German companies jointly. Thus they have control of all the larger North Atlantic companies, with the exception of the unimportant French Line and the very impor-

[a] There are numerous other clauses. See the reprint of this agreement in the appendix of the 1902 Report of the United States Commissioner of Navigation.

THE PORT OF HAMBURG

tant Cunard Line. The British Admiralty presented the latter with the "Mauretania" and the "Lusitania," on condition that it should remain outside the combine.[a] Its strong and independent position has been a thorn in the side of the International.

We have seen, then, the significance of direct steamship lines between a seaport and the oversea world. In proportion as Germany offered the prospect that such lines would be successful, they were established—often before the prospect was in sight. The old English middleman has been shaken off; the toll he took from the German producer and the German consumer has been done away with. The majority of the Hamburg lines, particularly those of the longer routes, are German; a large part of their securities is held in the Hansa city.

German emigration gave the Hapag the profits from which its fleet was built up. The prime significance of the present East European emigration that floods through Hamburg is that it furnishes the Atlantic liners with a west-bound human freight to fill their ships. The history of the Hapag is a fascinating story of the rise and growth of the largest and mightiest organization on the ocean today. It is part owner or partner of all but one of the large Hamburg steamship lines; in their union they are strong. Hamburg's line connections now reach to every continent, to every port of commercial importance. In versatility they are equalled by those of no other port in the world. The permanency and safety of each of Ham-

[a] 1902 Report of the United States Commissioner of Navigation, page 49. The terms of the loan made by the Admiralty to the Cunard Line and the new yearly subsidy granted, made the two steamers practically a gift.

HAMBURG'S OVERSEA LINES

burg's greater lines are assured by their close connection with each other, correlated as they are by the omnipresent Hamburg-American Line. Appearing as the Hapag does in international councils, with the united strength of the German lines behind it, it is the greatest force in ocean shipping today.

CHAPTER V

HAMBURG'S SHIPBUILDERS AND MERCHANTS
STATE AID TO THE MERCHANT MARINE

OTHER forces making for the present prosperity of Hamburg are its shipbuilding industry, the integrity, experience and activity of its merchants and the state aid which its steamship lines receive in the form of subsidies.

A recent occurrence in the United States illustrates the need of adequate docking and repairing facilities in a port. In April, 1910, the "Princess Irene" of the North German Lloyd ran aground off Long Island. She is one of the older and smaller vessels of the Lloyd and yet, when it was found necessary to have her repaired, she had to be towed to Newport News, the nearest port that had a dock sufficiently large to receive her. New York's excuse doubtless is that none of the large liners entering New York are domiciled there. Their foreign owners have all ordinary repairs, as well as the regular cleaning and overhauling, done in the home port; a New York dock built for the purpose of making extraordinary repairs could not support itself. However, docking facilities for the largest liners, if provided in one of our North Atlantic ports, would contribute considerably to its attractiveness in the eyes of foreign steamship companies looking for an American terminal.

The full efficiency of a merchant marine is only secured when there is at home ample opportunity for making all

SHIPBUILDERS AND MERCHANTS

ordinary and extraordinary repairs to the ships of that marine. Hamburg has two docks, each of them large enough to receive the largest steamer that enters the port, the "Kaiserin Auguste Victoria," 25,000 gross register tons. One, of 35,000 tons capacity, is owned by Blohm and Voss, the other by the Hamburg branch of the Stettiner Vulkan. There are numerous other yards with docking facilities, the largest being the Reiherstieg yard. They accommodate easily and swiftly all ships that call on them and are kept busy by the Hamburg merchant fleet. The Hamburg Stettiner Vulkan will have a dock ready for the 50,000-ton "Imperator," which it is now building for the Hamburg-American Line. A characteristic of the German yards is their preference for the floating dock over the masonry graving dock.

For various reasons the center of the German shipbuilding industry is shifting from the Baltic to the North Sea and particularly to Hamburg and Bremen. The following table indicates the change:[a]

REGIONAL DISTRIBUTION OF GERMAN SHIPBUILDING.

	Built on North Sea Gross Register Tons		Built on Baltic Sea Gross Register Tons	
	Steamers	Sailers	Steamers	Sailers
1889,	80,593	13,175	115,670	3,302
1907,	145,094	16,389	156,252	11,524

The change from wood to iron in the material for ships made it less important to be on the woody shores of the Baltic than to be near Rhineland and Westphalia, the centers of the iron and steel industry, in West Germany.

[a] Neumann, 50.

THE PORT OF HAMBURG

The Baltic shipbuilding ports, such as Dantzig and Stettin, lie some distance up navigable rivers and have no channels to the sea capable of floating the big liners if they build them; Stettin has a channel of seven meters (twenty-three feet) at low and high water: there is no tide in the Baltic. Hamburg has a channel of 32.8 feet at high water. When the Stettiner Vulkan built the "Kaiserin Auguste Victoria," she had to be propped up on scows before she could be floated down the Oder to the sea. This is the chief reason why the Vulkan has established a yard in Hamburg, now engaged in turning out the "Imperator."

Moreover, it is easier for yards on the North Sea to get workmen than for those on the Baltic. As a rule, the shipbuilding industry no longer trains its own apprentices. Most of its workmen come from the bridge construction works, boiler shops, engine works and machine shops inland, the shipyards getting the worse element from inland because they pay no higher wages, while they offer conditions far inferior as regards the health and safety of the workmen and regularity of employment. North Sea yards are nearer industrial West Germany, whence these workmen come. Finally, the establishment of yards in Hamburg and Bremen is encouraged by the amount of repair work to be done there. In slack times it is the repair work which keeps the yards employed, enables them to pay expenses and retain their workmen.

The first landmark in the history of modern German shipbuilding was the decision of the first German Minister of the Navy, von Stösch, in 1874, to have the navy built up in Germany and not abroad. The experience thus gained by the yards was valuable in teaching them naval

SHIPBUILDERS AND MERCHANTS

construction; today Germany builds a considerable number of naval craft for foreign purchasers, especially torpedo boats and destroyers. Yet this experience was of little avail in building merchant vessels. Foreign-built ships were admitted free to the German registry, while there was a tariff duty on raw materials for shipbuilding from abroad. German-made materials were so expensive, owing to the undeveloped state of the German steel industry, that ships built of them were far dearer than free ships imported from England. In 1879 the duty was taken off foreign shipbuilding materials.

Yet there was at first little tendency to patronize the German yards. The Hamburg-American Line and the North German Lloyd continued to buy their ships in England, with the exception of two small orders in the early eighties, placed in Germany by the Hamburg-American Line, each for a steamer of 3,500 tons. The real starting point in the building of merchant vessels was the subsidy which the German government gave the North German Lloyd in 1885. Lines to Australia and the Far East were subsidized, a condition being that the boats used should be built in German yards, of German material. This resulted in an order for six ships, costing ten million marks, being given to the Stettiner Vulkan. The Vulkan lost one and three fourth million marks in executing this first order for the Lloyd, but the ships were satisfactory. In 1889, after long deliberation, the Stettin yard was timidly given an order by the Hamburg-American Line for an express steamer, to cost five million marks, the "Auguste Victoria." Acceptance of the order was made conditional on the Vulkan finding banking houses which would guarantee to the Hamburg-American

THE PORT OF HAMBURG

Line repayment of the instalments it was to advance as the ship was constructed, provided she did not meet all requirements.[a] She was a great success.

After this preliminary venture, the Hapag continued to order some boats from Germany, for instance, the "Fürst Bismarck" and the "Deutschland." Yet the great patron of the German yards has been the North German Lloyd. Of the four hundred million marks it had spent for ships up to 1907, two hundred and sixty million marks had gone to German builders. The decisive change from English to German yards was made in 1892, when Wiegand became managing director for the Lloyd. The Lloyd's expenditures for new ships were as follows:[b]

EXPENDITURES OF THE NORTH GERMAN LLOYD FOR NEW SHIPS.

	To German Yards	To English Yards
	Million Marks	
1857-82,	1.5	68.5
1883-93,	36.3	53.7
1894-1907,	220.5	9.0

All the four famous express steamers of the Lloyd, the "Kaiser Wilhelm der Grosse," the "Kaiser Wilhelm II.," the "Kronprinz Wilhelm" and the "Kronprinzessin Cecilie," were built at the Stettiner Vulkan, in which the Lloyd has a financial interest. A few boats were built for the Lloyd by the Vulkan on the commission basis, such as Harland and Wolff of Belfast apply to their construction

[a] Denkschrift des Vulkans: "50 Jahre Schiffbaus 1857-1907," page 11.

[b] Denkschrift des Norddeutschen Lloyd, "50 Jahre der Entwicklung, 1857-1907."

SHIPBUILDERS AND MERCHANTS

work for the White Star Line, but this proved unsatisfactory and has been abandoned.

The Hapag publishes no such figures as those of the Lloyd, given above; if it did, the result would be different. Its policy has been first to have a liner constructed in England, usually at Harland and Wolff's, then to have it copied at a German yard. Such a pair were the "Amerika" and the "Kaiserin Auguste Victoria," the latter built by the Vulkan of Stettin. The value which the Belfast shipbuilders attach to their relations with the Hapag is apparent from the clause in their contract with the International Mercantile Marine Company, reserving the right to build for the Hapag at all times.[a] In very recent years the Hapag has begun to turn from this policy and build pairs of ships in Germany. For instance, the "Cleveland" and the "Cincinnati" were both built in Germany, one at the yard of Blohm and Voss and one at the yard of F. Schichau in Dantzig.

The other great German steamship lines, including those at Hamburg, build most of their ships in Germany. From an average of 35,000 gross register tons in the years 1873-79 German yards have risen to an output of 368,440 tons for the merchant marine in 1907.[b] Four hundred and thirty-five steamers of 311,103 tons and 516 sailing vessels of 57,337 tons were built. The following were the leading yards in construction in 1907:

[a] Harland and Wolff have a contract with the International Mercantile Marine Company, by which they agree to build for no other customer than the International, provided the latter can keep their yards busy. All construction and repair work for all lines, excepting the Hapag, is postponed if it interferes with carrying out the International's orders.

[b] Lloyd Zeitung: Die Fortschritte des deutschen Schiffbaus, page 39.

THE PORT OF HAMBURG

Chief German Shipyards, 1907.[a]

Yard	Number of Ships	Gross Register Tons
Flensburger Schiffbaugesellschaft,	10	42,504
Blohm and Voss, Hamburg,	5	42,110
Stettiner Vulkan,	7	42,105
Bremer Vulkan, Vegesack,	10	39,156
A. G. "Neptun," Rostock,	12	21,697
A. G. "Weser," Bremen,	8	18,006

Fifteen per cent of these 368,440 tons were built for foreign purchasers. It is true that 28 per cent of the accretions to the German registry in 1907 were bought abroad. Yet even this signifies a considerable degree of emancipation from the English shipbuilding industry.

German steelmakers, exposed to the competition of foreign shipbuilding materials since 1879, have so perfected their product that little English steel is now imported. German rolling mills long suffered because of the high railroad rates which they had to pay, under the normal tariff, in order to get their plates to the distant shipyard. English steelmakers had the advan-

[a] Nauticus, 1908, page 384. The following statistics, from the Germanic Lloyd's, show the activity of various shipbuilding centers at the close of the year 1909, and the position which Hamburg holds:

German Shipbuilding Centers, December, 1909.

City	Number of Yards	Vessels in Construction	Gross Tonnage
Hamburg,	13	155	81,000
Stettin,	4	78	72,000
Bremen,	2	28	45,000
Kiel,	4	66	39,000
Vegesack,	1	18	39,000

(Hamburg als Schiffahrts- und Industrieplatz.)

The "Fürst Bülow," of the Hamburg-American Line, about to be launched at the yard of the Vulkan, in Stettin.

SHIPBUILDERS AND MERCHANTS

tage of cheap ocean transportation to the German yards. In 1898 the Prussian state railways created an exceptional tariff for German shipbuilding materials: for long hauls a ton-kilometric rate of 1.2 pfennigs as against 2.2 pfennigs, the lowest rate for any of these materials under the normal tariff.[a] In 1907 the German shipyards ordered 137,000 tons of plates and 61,000 tons of profile steel. Of this quantity 26,000 tons of plates and 13,000 tons of profile steel came from England.[b]

The German shipbuilding industry employs normally about 65,000 men, 90,000 if one includes the workmen at the imperial naval yards.[c] The total of the stock issues of the private yards is one hundred and fifty million marks. It is estimated that German shipyards in the course of the last twenty-five years have bought German materials to the value of more than one and one half billion marks. We have already seen what a high position Hamburg occupies among shipbuilding centers. There are more than 10,000 men employed in that industry in the Free Port.

There are three principal ship types in the merchant marine: the pure freighter, the express steamer and the combined freight and passenger type. The pure freighter is either a tramp or a liner, the other two types belong solely to the regular lines. The typical freighter is of about 6,000 gross register tons and has a carrying capacity of about 8,000 tons. The entire ship is utilized for cargo, with the exception of crew quarters, engine

[a] Aftalion, 431. Railroad rates are discussed in the next chapter.
[b] Lloyd Zeitung: Die Fortschritte des deutschen Schiffbaus, pages 37-39.
[c] Ibid.

THE PORT OF HAMBURG

room and enough coal to carry the vessel eight to eleven knots an hour. The pure express steamer is designed to carry only passengers, their baggage, the mails and a very limited amount of valuable package freight. She has her entire carrying capacity full of cabins, steerage quarters, crew quarters, huge, powerful engines, boilers and bunker coal; she runs twenty knots and more per hour. Five hundred to one thousand tons is all the cargo one of these racers can carry. One of the best of the German express steamers is the "Deutschland" of the Hapag. She is of 16,500 register tons, carries no cargo, 800 cabin passengers, 300 in the steerage and a crew of 552. She runs at the rate of twenty-three knots an hour. These steamers are only profitable if they have a full passenger list, which limits them to the North Atlantic service. They count on postal receipts and a passenger patronage which desires speed and which accommodates itself to the sailing dates of the racers.

The combined type carries a great deal of cargo. Amidships, and particularly in the towering superstructure which this type of vessel carries, are the passenger accommodations. As compared with the freighter, the portion of her capacity which the ship can utilize for cargo is limited by the need of many rooms for elaborate passenger accommodations, larger quarters for the larger crews which the passenger steamer demands, larger engines and more bunker room to give the greater speed which passengers require. She runs twelve to twenty knots an hour. A typical example of the combined type is the "Pennsylvania" of the Hapag. She is of 13,333 register tons, carries 400 cabin passengers, 2,400 in the steerage, a crew of 200 and 14,300 tons of cargo.

SHIPBUILDERS AND MERCHANTS

If freight is scarce, such a steamer can make expenses by its passenger service alone; out of the passenger season, income from freight alone is profitable; when both freight and passengers are moving, a double profit is realized. Indeed, this is the principle of the combined type.

The Hapag and the Lloyd have pursued different policies in the ship types they have chosen. The Hamburg-American Line has built only one pure express steamer, the "Deutschland." She was put on in 1900 and held the "blue ribbon of the ocean" until the "Mauretania" and "Lusitania" came on the scene. The Lloyd, on the contrary, has on the North Atlantic service four steamers of over twenty-two knots; with them a weekly express service is maintained between New York and Bremerhaven. So in other parts of the world: the Lloyd has a postal service to the Far East, the Hapag a freight service. The Lloyd has forty-nine steamers, each of over twelve knots speed, the Hapag twenty-nine.[a] In 1907 the Lloyd received from the German and the American governments for carrying the mails between the two countries $422,154, the Hapag only $138,037.[b]

The passenger service of the Lloyd is probably still its chief source of income, as it was formerly the chief source of income for the Hapag as well. Nowadays the passenger trade of the Hamburg company pales beside its freight business. The greater number of fast steamers in the Lloyd fleet is an expression of the greater importance to it of the passenger traffic. The Lloyd is the

[a] 1910 Report of the United States Commissioner of Navigation.
[b] 1909 Report of the United States Commissioner of Navigation, page 39.

THE PORT OF HAMBURG

Cunard Line, the Hamburg-American the White Star Line of the German merchant marine.

Yet the Hapag by no means renounces the passenger trade. On the North Atlantic it meets the speed of the Lloyd express steamers by the luxury of its new combined types, such as the "Amerika," the "Kaiserin Auguste Victoria," the "Cleveland," the "Cincinnati," "Berlin," etc. On these boats, which have a speed of sixteen to eighteen knots, the cabins and social rooms are large and commodious, there are a number of suites with private dining and bath rooms, cafés where meals are served à la carte, gymnasiums, electric passenger elevators and wide promenade decks. In a brochure of the Hapag, the passenger is observed occupying himself as follows: "He betakes himself to the popular promenade decks and mingles with the throng of promenaders on this Atlantic Pall Mall or Rialto. He lingers here awhile to read the last reports from the stock exchange or hear of the latest occurrences on land, news of which has just been received by wireless. Then he goes to his luxurious cabin, taking his wife a bouquet of fresh-cut roses, to discuss with her whether they shall give a dinner that evening, in the Ritz-Carleton restaurant. By telephone he reserves the necessary places at table, orders a choice menu, invites his friends and receives their acceptances. After dinner they sit and drink coffee in the wonderful Winter Garden in the midst of palms and the odor of fresh flowers, listening to the strains of music."

A specialty of the Hapag has been the creation of third-class accommodation on the new combined boats. It was designed to give the more civilized German emigrant, at a slightly higher cost, better service than the Slavic

SHIPBUILDERS AND MERCHANTS

steerage passenger. The third-class passengers have their own cuisine and sleep in separate cabins instead of in huge rooms together, like the steerage people. Third-class passage costs on the "Kaiserin Auguste Victoria" and the "Amerika" 180 marks as against 160 marks for steerage.

Another factor in Hamburg's success has been the activity and integrity of its merchants. It was they who kept Hamburg's commercial relations intact with the Scandinavian and Baltic countries in the long interim between the days of the Hansa and the formation of the German Empire. It was they who, as the repression of the colonies of European powers relaxed, were quick to establish trade relations with those colonies, especially the dependencies of Spain and Portugal in South America and the West Indies. It was they who have stood behind the recent expansion of German trade to the Far East.

Through the centuries Hamburg has been attracting merchants of ability from all Europe. In 1189 Henry the Lion destroyed Bardowiek on the Elbe. Its merchants found refuge in Hamburg. When the Jews were expelled from Spain and Portugal, Hamburg offered them welcome.[a] It knew how to appreciate their money and their business ability. When the experienced Dutch merchants of Antwerp were driven from their homes by Spanish persecutions in the Netherlands, Hamburg received them. Secrets of the trade had a different meaning then than in these days of travel, books, atlases and government reports. When, during the same Spanish Wars, the Merchant Adventurers of London saw themselves con-

[a] Of the 100,000 inhabitants within the walls of Hamburg in 1804, 17,000 were Jews. (The Picture of Hamburg.)

THE PORT OF HAMBURG

strained to remove from Antwerp their staple place for central Europe, they were prevailed upon to come to Hamburg; they remained until after the Napoleonic Wars. In return for the privileges which they enjoyed in Hamburg, the Hamburg merchants were given special privileges in London, where their settlement was known as the Steelyard. When Catholic persecution drove the Huguenots from France, Hamburg opened its arms to this sturdy, industrious and prosperous stock.

The Hamburg merchant body has been reared in a hard school. They had to maintain their commercial connections through hundreds of years when Hamburg had no naval power and when the naval powers of other European countries were being ruthlessly used against trade rivals. An Englishman wrote in 1804: "Few cities can boast of so many mercantile houses of respectability, whether considered for their wealth or their integrity, and it is to their responsibility and steady sense of honor that Hamburg owes its present high position in the commercial world." They have retained the ancient Hanseatic custom of apprenticeship for their sons, who serve their terms in all parts of the world. They are found in the banking houses of London and New York, in shipyards of Belfast and Philadelphia, with grain merchants of Buenos Aires and Odessa, with cotton buyers of New Orleans and Galveston, with the exporters of Calcutta and Hongkong. They learn the language, customs and wants of the people with whom they live and come home a splendid body of men for the government and the business world to recruit from. The careful study which German merchants and manufacturers make of the wants of foreign customers, their honesty and the trustworthi-

SHIPBUILDERS AND MERCHANTS

ness of the goods they sell, are matters too well known to need discussion.

The seaport merchant is a middleman between two countries. It would seem as if this middleman might be dispensed with as postal service, catalogs, rating books, travel and personal acquaintance abroad bring buyers and sellers in different countries more closely together and make them better acquainted with each other's reliability. Seaport exporters and importers can be dispensed with in many dealings between Germany and the United States, England, France, Austria, Italy and other such well-known countries. The St. Louis wholesaler buys his cutlery direct from the factory in Solingen, the new cotton spinning mill in Chemnitz orders its machinery direct from the English manufacturer. But the factory in Solingen would hesitate to sell a native jobber in Guayaquil or Shanghai or Lorenzo Marquez or Trebizond. It prefers to sell to these places through the Hamburg exporter, whose business it is to follow the general conditions in each of these countries each year and to be acquainted with the reliability of each individual buyer. Business conditions and especially business connections in countries besides those in Europe, North America and Australia are not common knowledge. There are still secrets of trade.[a] The possession of these secrets causes many goods to be exported via Hamburg which, if they were sold from the factory, would find a nearer seaport.

Bulky exports such as sugar require concentration in

[a] Hamburg exporters do not view with unconcern the disclosure of these secrets. They object strenuously to the action of the Foreign Office in Berlin in informing German manufacturers of opportunities for sales abroad.

THE PORT OF HAMBURG

a market where they may be sold. Magdeburg, center of the beet sugar industry of Prussian Saxony, was formerly the great export market of Germany. Now the east of Germany has taken to the cultivation of the sugar beet and sends its sugar by water to Hamburg, which also receives all the export Bohemian sugar, brought down the Elbe on bonded barges. This is making Hamburg a sugar market that rivals Magdeburg; it attracts to the Elbe port many shipments of sugar formerly sent to Magdeburg. The existence of this market and the existence of Hamburg's steamship connections bring to Hamburg for export a great deal of East German sugar that would normally go via Stettin.

With regard to imports, the old monopoly of the seaport merchant has been broken, so far as bulky products of low specific value are concerned. There are two reasons why the grain markets have moved inland to Magdeburg and Berlin. One is the greater cheapness of waterway transportation as compared with the railway. This difference is particularly striking in the case of grain because the Prussian railways put a high tariff on imported grain (4.5 pfennigs per ton-kilometer), in order to protect the native grain of the agrarian interests that prevail in the Prussian Diet. This makes it advisable to take the grain in bulk far inland near the centers of consumption so that the rail charges for distribution may be as small as possible. The great grain elevators lie on the banks of the Elbe, not at Hamburg but at Magdeburg, Dresden and Berlin. Similarly in West Germany, which is still more dependent on foreign grain, the elevators and grain exchanges are at Duisburg, Düsseldorf, Frankfort on the Main and Mannheim. Hamburg and Rotter-

SHIPBUILDERS AND MERCHANTS

dam act primarily as transfer points for grain between the grain ships and the river barges. Moreover, the large per capita quantities of grain consumed by an industrial population inland make it possible to have several markets inland rather than one at the seaport.

It is different with the importation of "colonial wares," such as coffee, tea, cacao, rice, cane sugar, spices, tropical woods, natural dyes, tobacco and tropical hides. These goods are of such high specific value that considerable financial power is required to import them in bulk and carry them. Because of the small individual demand for these articles, it takes a wide hinterland to support a market. A dealer in goods in which quality plays so large a part must have a long and intimate acquaintanceship with his trade, and only the seaport merchants have this. For all these reasons Hamburg is a most important market for such products as coffee, cacao, tea, rice, spices and tropical woods. For coffee it is, after New York, the greatest market in the world. It deals primarily in Brazilian and West Indian coffee, Amsterdam being still the chief center for the East Indian product. Hamburg's position as a market is strengthened by the large transshipment trade in these wares to Scandinavia and Russia. It is apparent what an influence responsible, experienced, trusted importers can be in gaining for their port the trade of a large inland territory.

London still holds one advantage over its competitors. It is the financial center of the world. A Hamburg or New York importer of coffee pays with a draft on London. This not only gives the London banks liberal commissions; it gives them—and through them the English trade—an insight into foreign business relations. Moreover, London

THE PORT OF HAMBURG

is the great consignment market, the only European port to which all goods can be consigned. Often the foreign producer cannot wait to ship his goods until he has sold them. Often he cannot sell them afloat. So he consigns them to a merchant in London and draws on the merchant for a part of the value of the cargo. In London the consignee can always sell the cargo; if he wishes he can at once get money on it from the bank, on the basis of his warehouse certificate. Hamburg is a consignment market only for certain goods, such as coffee.

When one compares the oversea commercial position of Hamburg [a] with that of other seaports, one finds that the Hanseatic port is supreme in the transshipment trade to Russia and Scandinavia. Hamburg merchants occupy a leading position in South America and the West Indies. They are supreme in East Africa and, since Woermann in 1840 set out to gain a foothold on the West African coast, they have there a significance only equalled by Liverpool, whose domain West Africa has been since the old slave-trading days. Finally, trade with the Far East—China and Japan—has been intensively cultivated since the government subsidy to the Lloyd in 1885. In 1900, when the scramble in China began, Germany was quick to secure a commercial and naval base in Kiao Chao.

The German government does not hesitate to use its influence to further German trade. In 1910 the Roumanian state railways cancelled an order given for locomotives to a Belgian firm and transferred the order to German manufacturers. Some bitterness was felt in Belgium for it was believed that German official repre-

[a] See particularly Wiedenfeld, 293 seq.

sentations had had to do with the cancellation. It is still fresh in everyone's mind what a determined stand the German Kaiser took when he saw France about to absorb Morocco, as she had already absorbed the rest of northwestern Africa and monopolized its trade. Though Germany did not obtain all she desired at the Algeciras Conference, she did get the "open door" in Morocco. America has similarly insisted on the "open door" in Manchuria and the State Department in Washington announced that it was instrumental in securing the orders for the two Argentine battleships now being built in the United States.

There are many forms of subsidy for the merchant marine. Admiralty subsidies, such as those given by the British Admiralty to fast British steamers, are given the steamship companies in return for their holding their ships in readiness to serve as transports, scouts, etc., in time of war. There are shipbuilding and fitting-out subsidies such as are paid by France according to the register tonnage of ships built and according to the weight of engines installed. There are mileage subsidies, often graded according to the speed of the vessels, like those paid by France to her steamers and sailers, premiums so high that French sailing vessels have sailed the seas without cargo, come home and claimed a subsidy that made the voyage profitable.[a] A government may rebate to a steamship company its payments of the exceedingly high Suez Canal dues, as the Austrian government does. There are subsidies concealed in the mail pay given steam-

[a] In the Report of the British Royal Commission on Subsidies, reprinted in the 1903 Report of the United States Commissioner of Navigation, a case of this sort is detailed (page 282).

ship lines, when that mail pay is larger than would be given to foreign lines under the rates of the International Postal Convention. Where higher rates than these are explicitly paid, the amount of the subsidy is easily calculated. For instance, we pay American ships for carrying inland and foreign mails eighty cents per pound for letters and post cards, eight cents per pound for other articles. We pay foreign ships the international postal rate of thirty-five cents per pound for the former class of matter, four and one half cents for the latter. But when a lump sum is yearly paid to the carrier, as in the case of many English postal contracts, it is usually impossible to ascertain how much of the sum was fair pay for service rendered and how much was subsidy.[a]

The most important form of subsidy is that paid for regular sailings, with mail steamers of twelve knots speed or over, between the home country and foreign ports with which the existing volume of trade alone would not justify a direct connection. Various advantages are expected from such a line. There is better postal service with the country in question, there is more frequent and swifter connection with oversea colonies than would otherwise exist, which is desirable for administrative and military purposes. Above all, there is the expectation that new trade relations will be established because of the facilities which the new steamship line offers.

The German government pays no Admiralty, shipbuilding, outfitting or mileage subsidies. It pays for

[a] The German Postmaster-General declared in the Reichstag that English lines get from their government 70 per cent more than the German lines for the same service. In addition, the English steamers get Admiralty subsidies.

SHIPBUILDERS AND MERCHANTS

carrying the mails no rates higher than those of the International Postal Convention. Its subsidies are solely of the two following types: lump sum postal contracts and out-and-out subsidies—of which there are two—for the establishment of new lines. Four lump sum payments are made, on the basis of postal contracts. These are generally supposed [a] to be more than simple payment for mails carried. These payments are as follows:

GERMAN MAIL CONTRACTS, 1907.

COMPANY.	LINE.	PAYMENT IN 1907.
Jaluit Company, Hamburg.	Jaluit–Sidney–Hongkong.	120,000 Marks
Sartori Steamship Company, Kiel.	Kiel-Korsör.	178,000 Marks
Grand Ducal Postal Administration, Schwerin.	Warnemünde-Gedser.	80,000 Marks
J. Bräunlich, Stettin.	Sassnitz–Trelleborg.	320,000 Marks
		698,000 Marks

The last payment is for carrying the Swedish and Norwegian mails, the two preceding for the Danish mail. Other mail payments, made at the rates of the International Postal Convention, are as follows (principal items):

OTHER GERMAN MAIL PAYMENTS.

COMPANY.	LINE.	PAYMENT IN 1907.
North German Lloyd, Bremen.	Bremen–New York, Bremen–Brazil, Bremen–La Plata, Bremen–Cuba, Mediterranean–Levant, and other lines.	637,474[b] Marks

---

[a] Aftalion, 523.
[b] Of this, 627,305 marks for the Bremen–New York line.

THE PORT OF HAMBURG

COMPANY.	LINE.	PAYMENT IN 1907.
Hamburg-American Line, Hamburg.	Hamburg-New York, Hamburg-West Indies, Hamburg-Cuba-Mexico, Shanghai-Tsingtau-Tientsin, and other lines.	732,887[a] Marks
Hamburg South America Steamship Co., Hamburg.	Hamburg-La Plata, Hamburg-Brazil.	129,443 Marks
Woermann Line, Hamburg.	Hamburg-West Africa, Cape Town-Swakopmund.	173,865 Marks

The following are the two trade-route subsidies:

GERMAN SUBSIDIES, 1907.

COMPANY.	LINE.	SUBSIDY.
North German Lloyd, Bremen.	Bremen-Far East and Australia.	5,590,000[b] Marks
German East African Line, Hamburg.	Hamburg-East Africa and South Africa.	1,350,000 Marks

The first out-and-out subsidy was paid to the North German Lloyd. It was originally 4,400,000 marks, granted in 1885 for the establishment of a monthly service to Australia and a monthly service to China and Japan, both via Suez. In 1898 the China-Japan service was made semi-weekly and the subsidy raised to 5,590,000 marks. The Lloyd makes these regular sailings with boats of a speed between 12.2 and 14 knots, carries the mails free and government officials at

[a] Of this, 308,437 marks for the Hamburg-New York line.
[b] Since 1908 this subsidy is 5,820,000 marks. These figures from the 1909 Report of United States Commissioner of Navigation.

The new monster liner, the "Imperator" (50,000 tons), of the Hamburg-American Line, beside its express steamer, the "Deutschland."

SHIPBUILDERS AND MERCHANTS

a rate one fifth below the regular passenger rate.[a] The German government fixes the maximum freight and passenger rates which may be charged and puts strict limitations on the employment of Asiatic labor on the boats. The boats may not be sold without the permission of the government. The Lloyd is required to keep a separate account of its subsidized services. If the profit thereon exceeds 5 or 6 per cent, the German government is empowered to order the rates reduced or to require increased service from the steamship company.[b] As we have already learned, the subsidized steamers must be built in Germany, of German material. The subsidized Lloyd liners call at Hamburg—as well as at other European ports—before leaving for the East, which makes them as important to Hamburg as to Bremen.

A line of postal steamers to the Far East would never have been established without a subsidy. The extortionate Suez Canal dues (seven francs per register ton, which is 42,000 francs each way for a ship of 6,000 tons), as well as the fact that the English, French and Japanese postal lines to Australia and the Far East are subsidized, makes a subsidy imperative if Germany is to have direct connection with the East by means of fast steamers. The Peninsular and Oriental gets from the British government a subsidy of £330,000 for its lines to the Far East and Australia, or 6,600,000 marks as against 5,800,000 marks for the Lloyd. The subsidy for the Lloyd is calculated at

[a] This reduction is also allowed, out of courtesy, to foreign government officials. It may explain the partiality of English officials in the East for the Lloyd boats, a fact that Wiedenfeld ascribes solely to the superior efficiency of the German service.

[b] 1909 Report of United States Commissioner of Navigation, page 237.

THE PORT OF HAMBURG

5.05 marks per mile, while the Peninsular and Oriental steamers receive 6.8 marks per mile, those of the French Messageries Maritimes 8.8 marks per mile.[a]

It is is difficult to ascertain what part of the growth of Germany's exports to Australia, China and Japan is due to these subsidized steamers, the finest ships that go through the Suez Canal. When in 1898 the Lloyd's subsidy was increased to pay for doubling its service to China and Japan, Germany's exports to China, as compared with the average of the years 1881-85, had increased in weight as $6\frac{1}{2}$ to 1, in value as $4\frac{1}{2}$ to 1.[b] This movement has continued. From 1901 to 1907 Germany's exports to China and Japan increased as follows:[c]

GROWTH OF GERMAN EXPORTS TO THE FAR EAST.

To	Million Marks	
	1901	1907
China,	38	63
Japan,	67	102
	105	165

This is not the measure of Germany's exports to the Far East. A large part of West Germany exports via Antwerp, so that in German export statistics these goods appear as destined for Belgium. Hamburg has had a major part in those exports which appear in the export

[a] In 1907 the Lloyd lost money on its service to the Far East, but not on its Australian service. (Nauticus, 1908, page 377.)
[b] Schumacher in collection "Handels- und Machtpolitik," Stuttgart, 1900, Vol. II., page 225.
[c] Statistisches Jahrbuch für das Deutsche Reich.

SHIPBUILDERS AND MERCHANTS

statistics. Of the one hundred and sixty-five million marks exports sent from German ports to China and Japan in 1907, Hamburg sent one hundred and sixteen million. The Lloyd fills its ships largely at Hamburg.

The second out-and-out subsidy was given to the German East African Line, an ally of the Hamburg-American and Woermann Lines. The original subsidy of 900,000 marks was given to ensure direct connection between Germany and the east coast of Africa, a connection which the government desired for trade reasons and also in order to be in constant communication with the colony of German East Africa. In 1900 the subsidy was increased to 1,350,000 marks and the sailings of the company made correspondingly more frequent. Now the German East African Line despatches a steamer of ten to twelve knots speed semi-monthly from Hamburg to Africa, touching at Bremerhaven and various other European ports. These steamers circumnavigate Africa, alternately by the east and the west circuits. The company also has a monthly steamer plying on the east coast of Africa and maintains a service between the east coast and Bombay. The conditions attached to the granting of the subsidy are substantially the same as those for the Lloyd subsidy.

The subsequent growth of trade with East Africa is well evidenced by the following quotation from the Report of the British Royal Commission on Subsidies, rendered in 1902:[a]

"The German East African Line has made its way and obtained a firm footing, in spite of uphill work in the first years of its existence. The increase in value of purely

[a] Reprinted in the 1903 Report of the United States Commissioner of Navigation, pages 284 seq.

THE PORT OF HAMBURG

German goods traffic on the line rose from £300,900[a] in 1891 to £955,600 in 1898. With these figures may perhaps be compared the board of trade returns of British exports to Zanzibar and Pemba, which in 1892—the first year for which the figures are given—were valued at £105,670; in 1898 they were £114,217 and in 1901 £107,205. In this and almost every instance German money is so laid out as to bring the benefits of it to bear where German shipping is having a struggle, which, without assistance, would appear to offer little prospect of success. Great Britain, on the other hand, notwithstanding the area and growing importance of its East African territories and protectorates, has no fast, direct Imperial steamship communication with them, though since June, 1902, the British India Company has made an effort to run direct cargo boats; and our country has long been content with transshipment by British mail lines at Aden, with all the attendant losses by breakage, delay, weather, and native thefts, as well as loss by transfer of East African orders away from British manufacturers to their continental competitors.

"The center of the market at present tends to shift from London to the continent. The best mail service to the British center of government and the great emporium of Zanzibar is by foreign ships, and very occasionally even confidential official documents and British troops have been sent by the same means. The same is true of mails to British East Africa and Uganda, and British Central Africa. It seems of little use to have built the Uganda Railway at great cost, if the best means of communication

[a] An English pound sterling is $4.86; for rough reckoning, consider it $5.00.

SHIPBUILDERS AND MERCHANTS

with it and profit from it are to be left to foreign nations. An opportunity was missed to lay the foundation of a direct steamer line during the outward carriage of Uganda Railway material, as was elicited by Colonel Denny. The Messageries Maritimes started their line to Zanzibar in 1888; the German line started in 1890; the British India Company's steamers, which are not equal to those of the foreign companies, ran direct between 1889 and 1892, losing about £44,000 a year, which happens to coincide almost precisely with the original subsidy of £45,000 paid annually at first to the German East African Company. This was increased to £67,500 a year in 1900, in return for two separate direct services every four weeks, one on the east and the other on the west coast of Africa, meeting at Durban or some other port. The German steamers run at twelve knots; the British India to Aden run at nine, receiving £9,000 a year as mail subsidy. Your committee are informed that the Austrian Lloyd company has just decided to run a monthly service (and oftener if it pays) to East Africa, with steamers of 4,000 tons register and a speed of twelve to thirteen knots, chiefly for the export of Austrian goods. . . . . Your committee are of opinion that a special case exists for establishing direct British Imperial communication with East Africa through the Suez Canal by ships of up-to-date speed and accommodation, it being understood that any such subsidy should be granted for Imperial considerations."

However one may ascribe the growth of German foreign trade with these subsidized regions to the activity of German manufacturers and merchants rather than to the presence of German subsidized steamship lines, it is

THE PORT OF HAMBURG

certain that the German lines offer the manufacturers and merchants a far better channel to work through than lines from foreign ports would offer, with the heavy trans-shipment charges, delay, breakage, dependence on foreign agents, etc., which are attendant on such shipments. It is perfectly true that subsidized steamship lines cannot create trade when the right sort of merchant does not stand behind and support them. The heaviest subsidized merchant marines are not those that are growing most rapidly nor do they belong to the countries whose foreign trades show the largest increase. Huldermann[a] cites the following table from Lloyd's Register of British and Foreign Shipping for June 30, 1909:

SUBSIDIES OF VARIOUS MERCHANT MARINES.

Country	Merchant Marine Gross Register Tons	Subsidies Marks	Subsidy Per Gross Register Ton
England,	17,378,000	34,000,000	1.95 M
Austro-Hungary,	750,000	20,000,000	26.70
France,	1,894,000	53,000,000	28.00
Germany,	4,267,000	8,000,000	1.85
Italy,	1,320,000	16,000,000	12.10
Japan,	1,153,000	28,500,000	24.70
Russia,	972,000	11,000,000	11.30
Spain,	710,000	15,500,000	21.85

A surprisingly low rate of subsidy per ton signalizes England and Germany, which are the very countries whose merchant marines in the last ten years have shown the most remarkable growth. In this period England has increased her merchant fleet by five million tons, Germany has increased hers by two million tons. This is, of course,

[a] Page 39.

SHIPBUILDERS AND MERCHANTS

no argument against the subsidizing of individual lines by a country whose business men are capable of taking advantage of the opportunity provided them.

Other aids which are afforded to the German merchant marine and so contribute to the prosperity of Hamburg are the various differential tariffs on the Prussian state railways. One set of these differentials, those designed to promote German shipbuilding, has already been mentioned. The other rates will be discussed later. Two of the principal differentials, the prorating agreements with the German Levant Line and the German East African Line, might perhaps be properly treated under the head of subsidies. Others, such as the export rates for German manufactured goods, apply to all exports, whether by rail or water, and thus do not serve the particular interests of the merchant marine. Therefore railroad rates are reserved for discussion in a later chapter.

The German shipbuilding industry, then, competing against experienced English shipyards, has gradually won the confidence of native shipowners and has become a staunch support of the German merchant marine. The yards furnish prompt and adequate repair facilities for ships in German ports. They have also taken the initiative in producing for their merchant marine fine specimens of the two chief types of liner: the express steamer, such as the Lloyd uses on its North Atlantic service, and the combined freight and passenger steamers, such as the Hamburg-American Line prefers. Hamburg, because of its situation near the West German steel industry and because of the repair work that the port offers, has become a leading shipbuilding center; Hamburg is now building the "Imperator," the Hamburg-American Line's new

THE PORT OF HAMBURG

50,000-ton turbine steamer, which is to be the largest ship afloat when she is finished.[a]

Hamburg's merchants are a powerful influence in gaining for Hamburg's steamship lines exports and imports which would normally go via other ports, in enlarging Hamburg's dependent hinterland. They are personally an able body of men and inherit from their Hanseatic ancestors the valuable trade with Russia and Scandinavia. They have developed a heavy trade with South America and the West Indies, with the east and west coasts of Africa and with the Far East. Along with German manufacturers, they have developed a great thoroughness in studying the peoples and markets at which they aim. Their activities are primarily in exporting German products to the less cultured lands and in importing for Germany and the neighboring foreign countries colonial products. The Hamburg merchants are supported in their search for, and retention of, foreign markets by the power of the German Empire.

The only important German ship subsidies are those to the North German Lloyd for a service to Australia and the Far East and to the German East African Line for a service to Africa. Hamburg has the advantage of these subsidies because the German East African Line is a Hamburg company, while the Lloyd boats call at Hamburg and get the chief part of their cargo there. The establishment of both these lines has been attended with a large increase in German trade with the lands to which they run. That this increase in trade, however, is due in a high degreee to other factors than the existence of the

[a] Recently the yard of Blohm and Voss, in Hamburg, has begun construction of a sister ship to the "Imperator."

SHIPBUILDERS AND MERCHANTS

new lines, is apparent from the experience of other countries, notably France, whose enormous subsidies have not found at home merchants and manufacturers capable of filling the ships that are put in service. The German subsidies show careful investigation of trade possibilities before they were granted; their success testifies of that harmonious coöperation between government and individual, of the willingness and readiness of the latter to seize the opportunities which the former provides, which is characteristic of modern Germany. "Unsere Zukunft liegt auf dem Wasser."

CHAPTER VI

WATERWAYS AND RAILWAYS

CONSIDERED from the point of view of the seaport, its traffic by river with the interior is a sort of extended lighterage service. A port with a capable waterway behind it has an economical means of sending and receiving freight—not merely cheap, bulky freight—such as is lacking in a port dependent simply on railway connection with its hinterland. A port which is regularly despatching steamers three quarters loaded with the bulk goods that a river brings it, can offer most attractive ocean rates for rail-brought manufactured goods to complete the cargo. Moreover, not only does the port with an inland waterway get and send much of its freight more cheaply than it could by rail; the railroads, in order to prevent still more traffic from falling to the waterway, are compelled to offer especially low tariffs to retain it.

Hamburg is particularly fortunate in its waterway connections with its hinterland. The main rivers of Germany are the Memel (Niemen), Pregel, Vistula, Oder, Elbe, Weser and Rhine. Between the Weser and Rhine, and parallel to them, has been built the Dortmund-Ems Canal, from Dortmund to the German seaport Emden, designed to furnish a German outlet for the Westphalian coal and iron district. The Memel reaches the Baltic Sea near Memel, Königsberg is at the mouth of the Pregel, Dantzig at the mouth of the Vistula, Stettin at the mouth of the Oder. The Elbe flows into the North Sea at Hamburg, the Weser at Bremen, the Dortmund-Ems Canal at

WATERWAYS AND RAILWAYS

Emden; while the Rhine does service for the three foreign ports of Rotterdam, Antwerp and Amsterdam.

Thus there exists a complete system of waterways running from south to north. The necessary and natural thing was to complete these waterways by connecting links from east to west. In the case of the rivers from the Elbe eastward this has already been accomplished. The Memel, Pregel and Vistula are connected by a series of canals that skirt the seacoast. Eastern tributaries of the Oder, the Warthe and Netze, are navigable to a point so near the Vistula that it was simple to unite the two by means of the short Bromberg Canal. Similarly, the Havel, an eastern branch of the Elbe, has two canals to the Oder: the Finow Canal from Berlin to Hohensaathen on the lower Oder and the Spree-Oder Canal from the Spree, a tributary of the Havel, to a point near Fürstenberg on the Oder. The latter canal is intended to unite the Havel and the upper Oder. Hamburg, then, has waterway communication with all of eastern Germany, with Berlin and the Silesian industrial district, in addition to its own territory of the Elbe. This connection is denied Hamburg's rival, Bremen, for there is no canal between the Elbe and the Weser.[a]

[a] The Midland Canal will join Elbe and Rhine and so complete the national waterways system. It will take 600-ton barges. It was planned to run from Ruhrort on the Rhine to Herne on the Dortmund-Ems Canal, from Bevergern on the Dortmund-Ems Canal to Minden on the Weser, from Minden on the Weser to Magdeburg on the Elbe. The agrarians (Conservatives) in the Prussian Diet feared the cheap influx of foreign grain and, in the law which finally passed in 1905, broke the canal off at Hannover. As the canal is being built, the section Hanover-Magdeburg, is lacking. The same 1905 law provided for the construction of a 600-ton canal Berlin-Stettin, now under way.

THE PORT OF HAMBURG

A glance at the map shows that Stettin is the logical seaport for Berlin and Silesia. Silesia lies on the upper Oder, a stream taking 450-ton barges; Stettin is at its mouth. Stettin is 306 miles by water from Breslau, capital of Silesia; Hamburg is 490 miles. It is 160 miles from Berlin to Stettin via the Finow Canal; it is 240 miles from Berlin to Hamburg via the Havel and Elbe. Yet the rich province of Silesia and the city of Berlin, with its two million inhabitants,[a] are part of the hinterland of Hamburg. Four-hundred-and-fifty-ton barges reach Breslau from Hamburg as well as from Stettin,— often better, for conditions of navigation are better on the upper Oder than on the lower and boats can sometimes reach Hamburg more heavily loaded than they could have reached Stettin. Length of haul plays a minor rôle in internal navigation. There works in favor of Hamburg, and against Stettin, the attractive power of the great trade center which the Elbe port is. Silesia's sugar finds there a larger market, her exported iron and linen find better steamship connections with oversea, the needs of her industrial population find better stocked warehouses. The opening of the Kaiser Wilhelm Ship Canal, through the peninsular of Jutland, has made Hamburg a Baltic as well as a North Sea port.[b] Russian grain is brought to Hamburg at a rate per ton only one to two marks dearer than to Stettin, while the river rate from Hamburg to Berlin, via the 600-ton Elbe and Havel, is three marks cheaper than the rate Stettin-Berlin, via the

[a] Greater Berlin has three and one half millions.
[b] It was expected that this ship canal would give the German Baltic ports better access to the Atlantic. Instead, it has opened the Baltic to Hamburg and increased its dominance there.

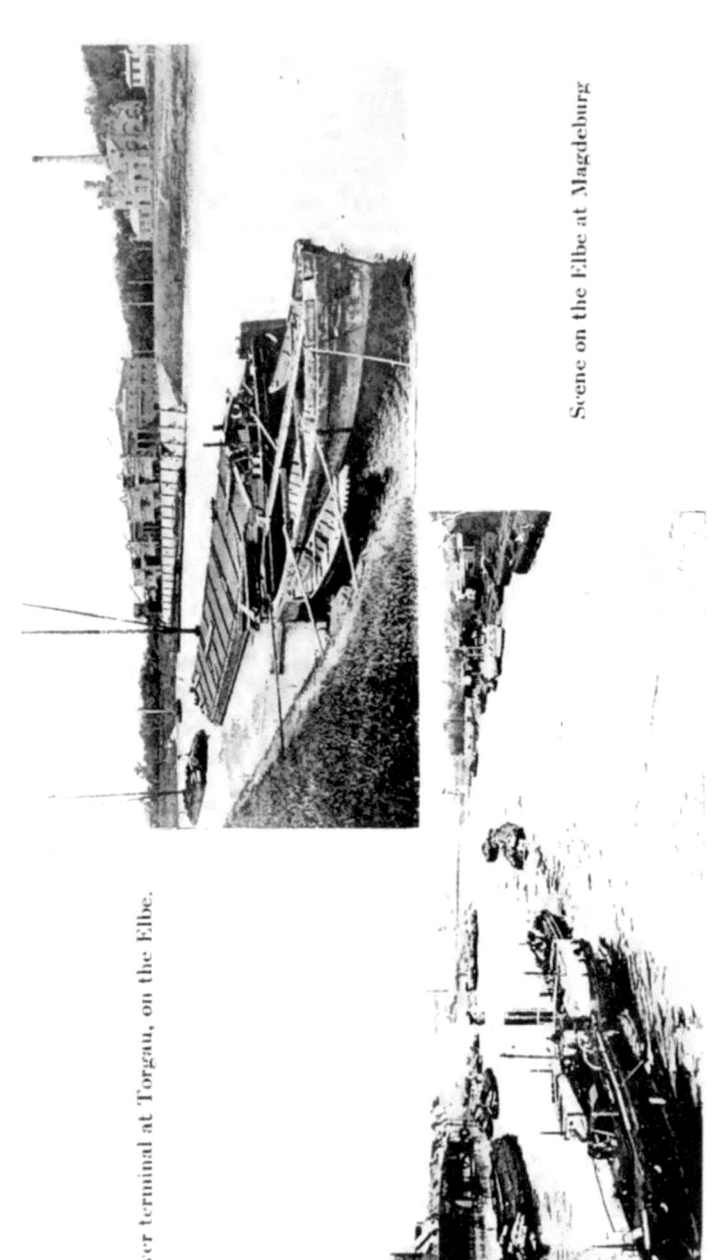

River terminal at Torgau, on the Elbe.

Scene on the Elbe at Magdeburg

Elbe river scenes.

antiquated Finow Canal, with its 170-ton barges.[a] The attractive power of Hamburg reaches even farther than the Oder. East Germany sends beet sugar to Hamburg for export, via the Netze and Warthe, Oder, Finow Canal, Havel and Elbe.

If Hamburg shares the foreign trade of Silesia and the east, it dominates the foreign trade of Berlin and the entire Elbe valley. The Elbe can be navigated with 800-ton barges from Hamburg to Melnik in Bohemia, the junction of the Elbe and the Moldau. The Moldau is similarly navigable from Melnik to Prague, 500 miles from Hamburg. The Havel, an eastern tributary of the Elbe, meeting it below Magdeburg and reaching to Berlin via the Spree, has already been mentioned. Above Magdeburg is a western tributary of the Elbe, the Saale, which is navigable for 350-ton barges to Halle. Proceeding up the Elbe from Hamburg we find such industrial centers as Magdeburg (190 miles) and Dresden (350 miles). Westward from Dresden is the Saxon industrial district (textiles and machines), the most thickly populated portion of Germany. Magdeburg is the center of the beet-sugar industry in the Empire; a few miles to the south, on the Saale, around Stassfurt, is the center of the famous potash fields. Back of Dessau is the center of the exporting chemical industry. It is a splendid hinterland and one bound up in oversea trade.

It must not be thought that the Elbe was always so efficient a waterway as it is today. All through the Middle Ages and until seventy-five years ago, the states bordering on the river were more interested in collecting tolls from passing ships than in furthering the interests of

[a] Aftalion, page 569.

navigation.[a] In the first half of the nineteenth century there were more than twenty-five toll stations on the Elbe. In those days little depth of channel was needed: the boats were small. Downstream they sailed or drifted; upstream they sailed or were towed by horses or men. It was a journey of several months from Saxony to Hamburg. Low water, which today does the damage, was less to be feared than high water, which overflowed the tow paths and often tied up the boats for weeks at a time. The towpath was the principal concern of the riverain state, in so far as it concerned itself at all about the well-being of the river. With the advent of the barges of today and the modern system of steam towage, interest shifted from the towpath to the channel.

Waterway improvement in Germany has always suffered under the disadvantage that it is the business of the states adjacent to the river and not of the Empire. The Elbe flows through the kingdom of Prussia, the duchy of Anhalt, the kingdom of Saxony and the Austrian province of Bohemia. Prussia has done the improvement work for Anhalt, but Saxony has lagged behind. Hanover, until it was absorbed by Prussia in 1866, did little to improve the condition of its portion of the river. In 1869 there were depths of only one and one half feet in the channel.[b] The work of improvement on the Elbe was done principally in the years 1880 to 1900.

There were dredgings and blastings to be carried out and cuts to be made. At curves revetments must protect the river bank. Where the river channel was too wide it

[a] Tolls on the Elbe were not abolished until the Elbe Navigation Act was signed by the riverain states in 1870.
[b] Buchheister, page 138.

WATERWAYS AND RAILWAYS

was narrowed, so that the new channel had a flow of water sufficient to keep itself cleaned out. Where the adjacent land was frequently overflowed, dikes parallel to the river were built, forming, with the river bed, a high water profile capable of carrying off the floods. This last was, to be sure, a regulation more to protect agricultural land from high water than to protect navigation from low water. Up to 1906 Prussia had spent on her portion (seven eighths) of the German Elbe forty-one million marks.[a] In the course of the last few years the Austrian government has canalized the Moldau from Prague to Melnik and the Elbe from Melnik to the German border, to a depth of two meters.

The result of these expenditures is that there is a low water depth on the Elbe as follows:[b]

LOW WATER DEPTH OF CHANNEL OF THE ELBE.

Hamburg-Magdeburg (190 miles) 1.16 meters=3 ft. 10 in.
Magdeburg-Bohemia (190 miles) .94 meter =3 ft. 1 in.
Bohemian Border-Prague
  (120 miles) (canalized) 2. meters=6 ft. 6 in.

By low water depth is meant that this is the average of the lowest water stage recorded in each of several successive years. Ordinarily barges can count on quite a different depth; Elbe barges with a capacity of 1,200 tons and a draught of six feet loaded are regular occurrences on the Elbe below Magdeburg.

The possibilities of further deepening the German Elbe

[a] Peters, II., 260. De Rousiers and Charles, page 106, state that the total sum expended on the Elbe above Hamburg in the years 1864-94 was one hundred and thirty-two million francs (one hundred and five and one half million marks).
[b] De Rousiers and Charles, 106 and 107.

THE PORT OF HAMBURG

channel, without canalization, are not great. The attainable depth at low water is dependent on the quantity of water then in the river. This cannot be confined to an indefinitely narrow, deep bed, for navigation needs not only depth of channel, but also a certain width. Shipping interests are united against the plan of canalization. They are not willing to sacrifice the speed and freedom of movement in an open river for the greater depth which a lock canal would afford.

As great as the change made in the condition of the river has been the improvement of the "rolling" or floating stock that operates on it. The old wooden barges have been displaced by barges of steel. In 1842 the largest Elbe barges carried 150 tons, in 1866 250 tons, 500 tons in 1877. Today the largest barges have a capacity of 1,200 tons. They are 230 feet long, 36 feet broad and draw about six feet loaded.[a] A 1,000-ton steel barge costs about 50,000 marks.[b] The advantages of the steel over the wooden barge are its greater durability—in spite of only a slightly higher original cost: 50,000 instead of 40,000 marks—its lighter draught for the same load, its slighter resistance to the current and its ability to better stand the rough treatment to which modern freight-handling machinery subjects it. First-class

[a] A German engineer gives the following dimensions for a typical Elbe barge: length over stem and stern posts, 256.8 feet; length on upper water line, 248.3 feet; length of the bottom, 232 feet; greatest breadth over the frames, 38.73 feet; greatest breadth of the bottom, 37.88 feet; least depth at the side, 6.54 feet; light draught, 1.32 feet; draught 5.9 feet when carrying 1,057 tons. (R. Blümeke at the Tenth International Congress of Navigation, Milan, 1905.—Proceedings of the Congress.)

[b] Thackera, 72.

barges are provided with officially approved devices for sealing their hatches; they are sealed and bonded for transportation between the Free Port and upper river points.[a]

In the old days barges were towed or sailed upstream and drifted downstream. As the wooden barge has given way to the steel barge, so the horse on the towpath has given way to the tugboat. When steamboats were first introduced, no one thought they would displace the barge and towpath service for freight. The first steamers were passenger carriers. Their engines were so heavy and they used so much coal that if much freight was carried the boat could not be kept afloat in the shallow channel. It was from Holland that the happy idea came of towing barges; the Dutch first applied it on the Rhine. Where the draught is limited, a towboat loaded down with engines can pull upstream in a line of barges more freight than it could carry if half loaded with engines, half with freight. On the same principle, a strong horse can pull in a cart behind him more weight than a weak horse can carry on his back. The best form of towboat for the Elbe is a low, shallow-going, side-wheel tug.[b] The largest have 1,200 indicated horse power, a length of 260 feet, a width of 29 feet (61 feet over the paddle boxes) and a

[a] Bulk cargoes, such as grain, are usually sent inland on bonded barges. In the river port the barge is unsealed and unloaded in the presence of a customs official, who assesses the duty. Coffee, cacao, etc., destined for storage in a river port, are sent up on bonded barges and stored in bonded warehouses.

[b] There is a towing chain which lies on the bed of the Elbe from Niegripp to Melnik (285 miles). Chain towing steamers still propel themselves upstream by winding this chain around a drum; most of them have been displaced by the side-wheel tugs which operate more economically.

THE PORT OF HAMBURG

draught of five feet.[a] Such a steamer carries a crew of twelve men and costs 300,000 marks. The tugboat can take 6,000 tons of cargo upstream in a tow of eight to ten barges.[b] A tow of 6,000 tons is equivalent to 600 German ten-ton freight cars or twelve freight trains.

Now that the channel of the Elbe has been deepened and light, powerful compound engines have been invented, it is possible for express steamers to carry freight. The largest of these have 350 horse power, are 213 feet long, 22 feet wide (42 feet over the paddle boxes), and draw five to six feet loaded. They have their own tackle for loading and unloading, which they use at landing places where there are no pier cranes. In all Elbe river harbors, these express steamers are given preference over all other craft, in assigning berths and pier cranes. They are considerably used, not only for trade between inland points and Hamburg, but also for purely internal services: Magdeburg-Berlin, Magdeburg-Breslau, Berlin-Breslau, etc.[c]

[a] Thackera, page 73.
[b] Barges are also towed downstream even when empty, in order to get another cargo more quickly.
[c] The United Elbe Navigation Company, the largest Elbe company, has the following regular services, performed by steamers running from two to four times per week:

> *Express Services of the United Elbe Navigation Company.*
> Hamburg-Magdeburg and vice versa.
> Hamburg-Halle and vice versa.
> Hamburg-Riesa and vice versa.
> Hamburg-Dresden and vice versa.
> Hamburg-Laube-Tetschen and vice versa.
> Magdeburg-Riesa-Dresden and vice versa.
> Magdeburg-Laube-Tetschen and vice versa.
> Dresden-Riesa to Magdeburg and Lübeck.
> (Vereinigte Elbschiffahrts-, page 4.)

WATERWAYS AND RAILWAYS

Besides these regular types of Elbe vessel—the barge, tugboat and express steamer—there are two special types that deserve mention: the beer barges and the oil barges. The United Elbe Navigation Company maintains a semi-weekly service with twelve beer barges from Laube-Tetschen in Bohemia to Hamburg, calling at Riesa in Saxony, to bring beer for export, principally to America. The Pilsener beer is transshipped from rail to barge at Laube-Tetschen, the Münchener at Riesa. These beer barges are simply built-over barges, supplied each with pipes and a cooling machine. The beer is thus transported more cheaply than by rail, at a more even temperature and with less jolting.[a] The oil barges are put in service by the German American Petroleum Company, subsidiary of the Standard Oil Company, and by the Pure Oil Company. The barges have capacities up to 1,000 tons and are divided into compartments (like the oil steamers) for different kinds of oil. The transportation of oil is one unbroken flow from well to consumer. The oil flows from the wells in America to refineries on the Atlantic coast, is refined and pumped into tank steamers. In Hamburg it is pumped into the river barges. Each Elbe harbor has its oil tanks on the river bank, into which the oil is pumped, there to remain until it is sent inland by tank cars. The barges, harbor tanks, tank cars, the city oil wagon and the can with which the driver measures out his ten litres of oil to the inland grocer, are all property of the German American Petroleum Company. No money is wasted on middlemen.

The organization of transportation has also expe-

[a] According to the statistics, Laube in 1907 sent 6,581 tons of beer downstream, Riesa 5,156 tons.

THE PORT OF HAMBURG

rienced a modernization: the independent boatman, owning his little barge, is being displaced by the large companies, whose boatmen are employees. The large companies own tugboats as well as barges, they have representatives in the river ports and agreements with the forwarders inland, they make long-time contracts with inland and riverain shippers, they have agreements with transoceanic steamship companies and give through bills of lading to foreign points. Only the companies can afford to own or lease terminal facilities in the river harbors.[a] All these factors work in their favor against the independent barge owners, who are dependent for freight on casual offers which they get on the Shippers Exchange of Aussig, Hamburg or Magdeburg, or from individuals, and who are dependent for towage on the tugs which the companies own. The express goods business is of course entirely in the hands of the companies, who alone can own express steamers. The cream of the business in bulk goods, on the Elbe, falls to the companies; in the service Hamburg-Berlin and on the waterways of the Mark Brandenburg (Havel, Spree, Finow Canal, etc.) it belongs primarily to the independents. The independent and his family live on the barge and he need earn only enough to pay for their food and clothing. On all German waterways there is a great superfluity of barges, which keeps freight rates down.

This superfluity of barge room on the Elbe and the possibility of at any time calling in barges from the Mark or the Oder make it impossible for the companies to com-

[a] However, most river terminals are owned and operated by river cities at rates that are uniform for companies and independents.

WATERWAYS AND RAILWAYS

bine and raise rates. In 1904 three of the main Elbe companies were merged into the United Elbe Navigation Company (Vereinigte Elbschiffahrts-Gesellschaften A. G.) with a capital of eleven million marks. The success of the merger emboldened the United in 1906 to lease for ten years the properties of the two remaining companies, one of them an association of independents.[a] The result was the formation of a new formidable rival (the Neue deutsch-böhmische), which, followed by the slump in business that began in 1907, made it difficult for the United to keep its head above water.

The United Elbe Navigation Company, employing 6,000 persons and doing a larger part of the transportation on the Elbe than any other company,[b] has 103 river and 41 harbor tugboats and 1,200 steel barges[c] of 750,000 tons carrying capacity, 19 express steamers, 250 lighters and 75 auxiliary vessels, such as launches, floating cranes, storage barges, etc. It has a large number of warehouses in the upper Elbe ports, besides terminal facilities and

[a] The value of the property of all the controlled companies is thirty-one million marks. (Vereinigte, page 12.)

[b] The traffic of the company is shown by the following table:

Freight Moved by United Elbe Navigation Company from Elbe Stations.

	Downstream	Upstream
1907	1,048,461	1,829,055
1909	2,204,679	2,152,790

(Vereinigte, page 6.)

Of these quantities, there reached Hamburg in 1907, 823,339 tons, there came from Hamburg 1,658,992 tons. The United has its own shipyard in Dresden.

[c] Thackera, page 72.

equipment at Hamburg, of which we shall speak later. It has a horde of agents[a] in river ports and arrangements with forwarders, inland from the Elbe, to collect freight; it has a contract with the Hamburg-American Line by which it gives a through bill of lading from the Elbe river port to inland points in the United States. The United operates principally on the Elbe; on the Havel and the Mark waterways the leading companies—interested primarily in the express service—are the Berliner Lloyd and F. Andreae of Magdeburg.

The terminal facilities for river barges in Hamburg have already been partially described. Imported English coal is discharged into chutes leading to the barge, by baskets which the ship's tackle hoists. Grain is sucked out of the hold of the ship through the long proboscis of the floating grain discharger, cleaned, weighed and slid into the barge, the discharger lying between it and the ship, made fast to a mooring post. Incoming liners tie up at their berths and their cargo is rushed ashore by cranes or dropped overside by their own tackle into waiting lighters and barges. As has been explained, a barge is allowed to come into the seaship basin and lie alongside the ship only when there are at least fifty tons to be transferred between the two. Otherwise the barge must wait in one of the barge basins while a lighter brings to it the small charge from the ship. As the liner's chief care is to discharge her cargo as rapidly as possible,

[a] The United Elbe Navigation Company has offices in Prague, Melnik, Aussig, Rosawitz, Bodenbach and Tetschen, in Bohemia; in Schandau, Dresden, Meissen, Riesa, and thirty-one other points in Germany.

WATERWAYS AND RAILWAYS

many goods discharged into the quay sheds are destined for shipment inland on the Elbe. Barge or lighter calls for these goods when the liner has left her berth.

The United Elbe Navigation Company has in the Moldau barge basin at Hamburg a new and ideal pier shed for assembling and distributing barge cargoes of package freight. The shed is built over the water so that lighters can be towed under the center of it. Six arriving barges can lie along its outer edge, 1,450 feet long, and be unloaded by eighteen portal cranes, the goods being trucked inside the shed and sorted. Then they are, by means of hand windlasses, let down through the hatchways in the floor to the lighters waiting below, and by these lighters carried to various ships which await them. The shed has 86,000 square feet storage surface and was built by the Hamburg state for one and one fourth million marks. The United pays a yearly rental of 100,000 marks therefor. To do this distributing and collecting by lighters the United operates in Hamburg, as has been seen, 250 lighters and 41 harbor tugs for them.

There must be water space for the swarm of barges waiting for their ships to arrive, waiting for a tow to form and take them upstream, or waiting for orders. These barge basins are strung behind the fan-shaped collection of seaship basins on the left bank of the river. They communicate with each other and each has an opening into the closed end of the seaship basin behind which it lies. So a barge can slip into the basin where its ship is loading and slip out again without interfering with the heavy traffic on the main stream. The barge basins are connected by a cut with the Elbe above Hamburg, so

THE PORT OF HAMBURG

that the barge traffic occasions the least possible disturbance of the ocean vessels.[a]

Good barge terminals are just as necessary in the upper Elbe harbors, not, as in Hamburg, in order to facilitate the exchange between two water carriers, but to offer smooth transshipment between barge and freight cars. The proper function of a river port is not primarily to serve as destination or source of cargoes but to act as transshipper of cargoes from, and destined for, the inland. This collecting and distributing must be done by the railroads. In order that this may be done, two things are necessary. There must be physical connection between the two carriers: the opportunity of effecting the exchange at slight expense. Otherwise it will pay to make shipments between Hamburg and inland entirely by rail. Secondly, the railroads must not apply tariffs for their distributing and collecting services which are out of all proportion to their tariffs for direct rail service between Hamburg and inland points.

A good river harbor needs water space enough so that many barges can simultaneously be loaded, unloaded and towed about without mutual interference.[b] The bed of the harbor basin, slanting upstream and cut into the land as the basins of the seaport are, must be below the bed of the stream outside. Perpendicular quay walls allow the steamers and barges to come alongside within reach of the cranes that top the wall. Transshipping machinery stands on the edge of the quay wall and

[a] For area of the barge basins and their proportion of the whole harbor area, see page 59.

[b] Dues from craft using the harbor for winter quarters are a good source of revenue.

WATERWAYS AND RAILWAYS

shortens as much as possible the unprofitable time spent loading or discharging. Cranes are used for package freight; grain is scooped out of the barges and conveyed into the elevators that lie directly on the river bank. Coal is discharged by traveling coal hoists similar to those in use on our Great Lakes.

Such a harbor must have a double set of railway tracks on the quay wall affording opportunity for direct transfer between car and barge. Behind the tracks, but within reach of the swinging cranes, are sheds to shelter the goods while they are being sorted to be drayed to the city or sent inland by rail. Customs officials must be at hand to assay the dutiable foreign wares that arrive in bonded barges. Local merchants require at the waterside a municipal bonded warehouse under control of customs officials, where imported coffee, cacao, etc., can be kept, the duty for them not to be paid until they are removed. The grain trade requires enormous elevators for imported grain, standing at the waterside and, by means of the "legs" that they let down, filling themselves directly from the barges. The petroleum which arrives demands tanks in the harbor with piping through which the oil is pumped into them from the oil barges, thence to be sent inland by tank car or distributed in the river city by the oil cart.

In Prussia, the harbors are as a rule built and operated by the cities. So with the harbor belt line; the city owns and operates it and hands the cars, properly classified, over to the state railways for further shipment. The municipal harbor at Magdeburg cost 8.2 million marks.[a]

[a] Die städtischen Hafenanlagen in Magdeburg, Magdeburg, 1898, page 53. The municipal harbors in Germany rarely charge dues sufficiently high to make a profitable return on the money invested

THE PORT OF HAMBURG

The harbor at Dresden cost seven million marks. It is rectangular in shape, has a length of 4,000 feet and a breadth of 400 feet. The harbor entrance is 115 feet wide, the water area 37 acres.

The Prussian state railways have not that feeling of community of interest with the waterway which is so lauded in America. The railway, by willingly acting as feeder and distributor for the waterway, condemns itself to unprofitable short hauls. There is money in hauling Indian cotton, coffee and rice from Hamburg to Leipzig at the special import rates accorded these materials. There is none in hauling these articles at these reduced rates from Torgau, a transshipment harbor on the Elbe,[a] to Leipzig. Similarly, if the export rates allowed Leipzig manufacturers for railway shipment Leipzig-Hamburg direct, were allowed for the short haul Leipzig-Torgau, these exports would seek the river port. The policy of the Prussian railways is discriminatory against the waterways in that the special import and export rates accorded goods

in them. They are prevented by their mutual competition from putting their dues high enough to attain this. In 1898 Magdeburg's net operating income was 225,000 marks, less than 3 per cent on the investment, to say nothing of not providing for maintenance and depreciation. Conditions are no doubt the same today, if one can judge from the analogy of harbors on the Rhine. In 1907-08 Cologne and Düsseldorf each had large deficits in their harbor budgets. Düsseldorf lacked 400,000 marks of getting a proper return on the harbor investment of eighteen million marks. Yet this burden is cheerfully borne. The river port brings to the river town so much business and prosperity that the deficit is more than covered by increases in the income and property taxes.

[a] Therefore cotton, coffee, rice, etc., which enjoy import rates by rail from Hamburg, are transported inland from Elbe points at the customary inland rates.

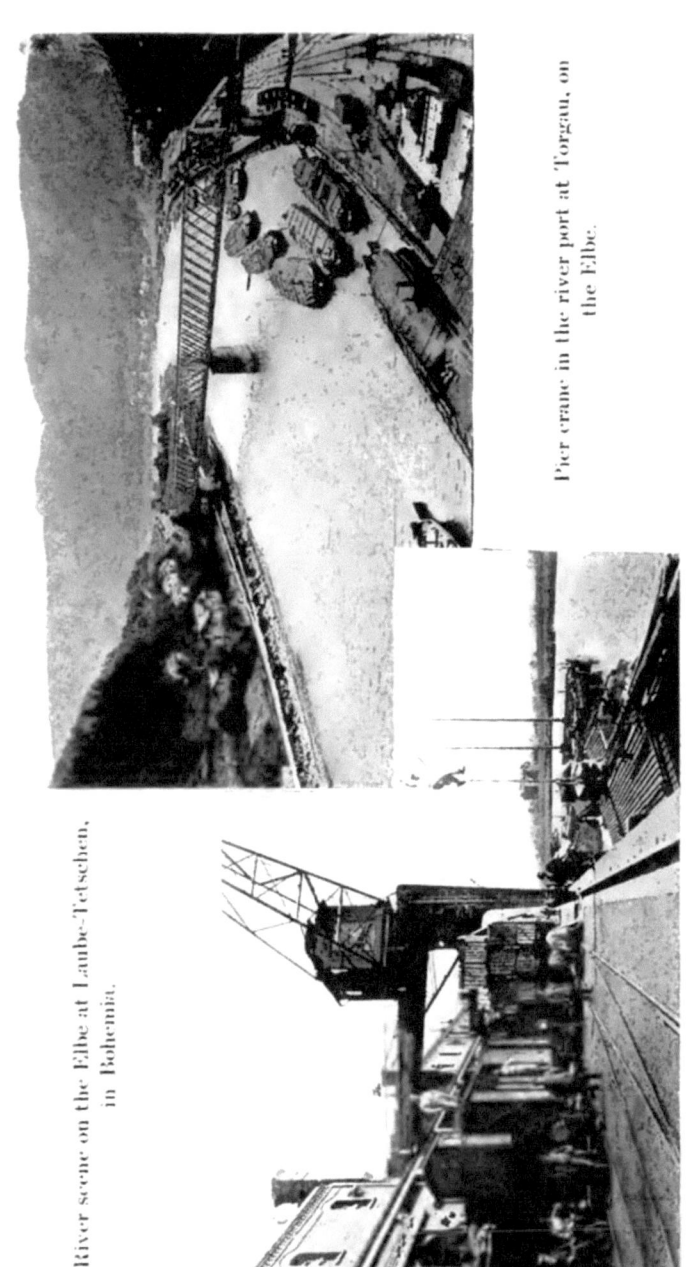

Pier crane in the river port at Torgau, on the Elbe.

River scene on the Elbe at Laube-Tetschen, in Bohemia.

Elbe river scenes.

WATERWAYS AND RAILWAYS

that use the railway for their entire journey are denied to those that use a waterway for part of their way; these must be carried at the higher "normal" rates. The discrimination, however, is not severe and—particularly on the Rhine—industrial interests have caused several exceptions to this policy. The one exception in the German Elbe territory is the extension of the export rate for sugar to shipments via Elbe ports, made in 1901.[a] This extension still endures.

However, the Elbe finds more coöperation on the part of the Saxon and Austrian railways. If freight goes as far as Magdeburg or Torgau by water, it means a short haul inland for the Prussian railways instead of the long haul from Hamburg. On the contrary, it is a matter of indifference to the Saxon railways, as far as length of haul is concerned, whether they take freight from the Prussian railways at the Saxon border or whether they take it from the river craft at Riesa or Dresden. It is similarly indifferent to the Austrian railways whether they begin their hauls at some railway border point or at Laube-Tetschen, a Bohemian Elbe port just across the border from Germany. Therefore the Saxon and Austrian railways make rate concessions to goods arriving by the Elbe similar to those they make for goods they take over from the Prussian railways. As a result of this coöperation on the part of the railways inland from the upper Elbe, the river's tributary territory is greatly extended. Austria exports via the Elbe more than via its own chief seaport, Trieste.[b] In 1907 there crossed the

[a] Cords, page 86.
[b] Major Placke, at the Erster deutscher Binnenschiffahrtstag, 1909.

THE PORT OF HAMBURG

Austrian border on the Elbe 633,800 tons,[a] mostly from Hamburg. Bremen has only rail connection with Austro-Hungary, to which it exported 84,500 tons in 1907.[b] This illustrates what the Elbe means to Hamburg in the competition for the German and foreign hinterland.

There is some difficulty in ascertaining what the water rates really are. The only ones which are published are the daily quotations of rates for a few bulky goods as made on the Shipping Exchanges of Hamburg, Magdeburg, etc. These rates vary enormously with the water stage and the demand for barge room. The best we can do is to take the average of them and say that this is the rate for water transportation. The grain rate from Hamburg to Berlin (225 miles) was on the average, for the years 1900-05, 3.57 marks or $.85 per ton. This was a ton-mile rate of $.0038. Sugar from Breslau to Hamburg (490 miles) in the years 1897-1906 was carried at the average rate of 5.69 marks or $1.35; the ton-mile rate was $.00276.[c] The corresponding rail rates are as follows: grain Hamburg-Berlin (173 miles) 13.75 marks or $3.01;[d] sugar Breslau-Hamburg (374 miles) 11.04 marks or $2.63.

There are two objections to water freight rate exhibits of this sort. In the first place, the water rate does not include insurance. In the second place, what we want to know is not the rate by water from the single city of

[a] Die Binnenschiffahrt in Jahre 1907 (Reichsstatistik).
[b] Statistik der Güterbewegung auf deutschen Eisenbahnen, 1907.
[c] Peters, II., pages 165-6.
[d] The high grain rates charged on German railways for imported grain are designed to protect native grain from the competition of foreign.

WATERWAYS AND RAILWAYS

Breslau to Hamburg but the rate to Hamburg by rail and water from various Silesian points not directly on the Oder. The majority of the shipments on German waterways are collected from inland. The rate comparisons we want, therefore, are those for water-and-rail as against pure rail rates. These combined rates are not published but are carefully kept secret by the inland shippers and the transportation companies, out of fear that the state railways, on learning them, will institute competitive rates. A fortunate circumstance brought some of them to light. In 1904 the Elbe dried up in the summer season and various inland industries dependent on river transportation saw themselves threatened with destruction. Chambers of commerce petitioned the Prussian railways temporarily to grant tariffs as low as water transportation had been. Statements of combined rates were appended to the petition. Some of them follow:[a]

COMBINED WATER-AND-RAIL RATES.

I. Oil from Hamburg to Cottbus.

(Ten-ton carload shipments.)

		Per Ton
A.	By rail direct (260 miles),	25.30 Marks=$6.02
B.	With transshipment from water to rail at Riesa,	
	Water freight Hamburg-Riesa (320 miles),	4.00 Marks
	Insurance,	1.00 Mark
	Transshipment costs,	.80 Mark
	Rail freight Riesa-Cottbus (63 miles),	7.30 Marks
	Rate for combined route (383 miles),	13.10 Marks=$3.12
	Difference in favor of combined route,	$2.90

[a] Peters, II., page 157.

THE PORT OF HAMBURG

II. Grain from Hamburg to Cottbus.

(Ten-ton carload shipments.)

		Per Ton
A.	By rail direct (260 miles),	19.20 Marks=$4.57
B.	With transshipment from water to rail at Riesa,	
	Water freight Hamburg-Riesa (320 miles),	3.00 Marks
	Insurance,	.30 Mark
	Transshipment,	.70 Mark
	Rail freight Riesa-Cottbus (63 miles),	5.70 Marks
	Rate for combined route (383 miles),	9.70 Marks=$2.31
	Difference in favor of combined route,	$2.26

III. Ground Glass, Cottbus to Hamburg.

(Ten-ton carload shipments.)

		Per Ton
A.	By rail direct (260 miles),	15.20 Marks=$3.62
B.	With transshipment from rail to water at Riesa,	
	Rail freight Cottbus-Riesa (63 miles),	4.70 Marks
	Transshipment,	.80 Mark
	Insurance,	.30 Mark
	Water freight Riesa-Hamburg (320 miles),	4.00 Marks
	Rate for combined route (383 miles),	9.80 Marks=$2.33
	Difference in favor of combined route,	$1.29

Examples in which the various items of expense in the combined route are reckoned still more in detail are the following:[a]

[a] The following two examples were kindly furnished by the United Elbe Navigation Company.

WATERWAYS AND RAILWAYS

East Indian Cotton, Hamburg-Chemnitz.

(*Ten-ton carload shipments.*)

		Per Ton
A.	By rail direct, Hamburg-Chemnitz (283 miles),	21.70 Marks
	Terminal charges in Hamburg,	1.50 Marks
		23.20 Marks=$5.76
B.	With transshipment, water to rail, Riesa,	
	Lightering in Hamburg,	1.80 Marks
	Transshipment, lighter to barge,	.50 Mark
	Water freight Hamburg-Riesa (320 miles),	5.00 Marks
	Transshipment in Riesa, barge to car,	.90 Mark
	Rail rate Riesa-Chemnitz (44 miles),	4.10 Marks
	Rate for combined route (364 miles),	12.30 Marks=$2.93
	Difference in favor of combined route,	$2.83

Munich Beer for Export.

(*Ten-ton carload shipments.*)

A. Direct by rail, Munich-Hamburg (500 miles), 45.00 Marks=$10.71
B. By rail Munich-Riesa (325 miles), thence by water to Hamburg (320 miles), in all, ... 35.80 Marks=$ 8.52

Difference in favor of combined route, $ 1.19

Of course the Elbe does not send many goods to Munich or get many from there, but the instance given illustrates how far out, for certain articles, the outer edge of Hamburg's tributary territory lies.

Hamburg has no waterway connection with South and West Germany; for trade with these regions it is dependent on the railways and the rates they give. The importance of the tariffs of railways serving a seaport lies in

THE PORT OF HAMBURG

their power to attach to the port trade which might "naturally" go elsewhere. The choice of a seaport depends on the rates between it and competitive territory, not on the greater nearness in space and time which one of the competitors has to its advantage. The need in German seaports for aid from the state railways is apparent when we consider the situation of Germany. The southern part of the Empire gravitates towards the Mediterranean ports of Marseilles, Genoa and Trieste. The Rheinish-Westphalian industrial district in the west and the entire rich Rhine valley trade naturally not with Hamburg and Bremen but with the Rhine seaports, Rotterdam, Antwerp and Amsterdam. Private railways pursue their own pecuniary profits; state railways can take account of the larger interests of the land they serve. They have done a great deal to prevent West Germany from becoming entirely dependent on foreign seaports.

German railways are operated by the various states,[a] the principal system being the Prussian-Hessian. Prussia has a preponderating influence in the Conference which determines the joint rates, and the ton-kilometer[b] rates of other states are so modeled on the Prussian that, for practical purposes, an examination of the Prussian tariffs serves for all. Prussian freight rates (the "normal" tariffs) consist of a system of ton-kilometric rates for different classifications of goods and for fast and slow service, the rate in some cases decreasing with the length

[a] There are a few railways, none of them main lines, still in private hands.

[b] A kilometer is .621 mile; 5-8 mile. The ton-kilometric rate multiplied by 2-5 gives the ton-mile rate.

WATERWAYS AND RAILWAYS

of haul.[a] The tariff for any particular haul is ascertained by adding the sum attained by the ton-kilometer calculation to a lump despatching fee, varying from 60 pfennigs to 2 marks per ton according to the class of the goods. Reduction in the cost of transportation of any article can be effected by "de-classifying" it—reducing it to a lower, cheaper class—or by creating for it a new "exceptional" tariff; reduction in the ton-kilometric rate is accompanied by a corresponding reduction in the despatching fee. The number and importance of these exceptional tariffs for carload shipments is so great that under them for years have been moved over 60 per cent of all traffic on German railways.[b] The "normal" tariff

[a] The rate per ton-kilometer for L. C. L. shipments is graded for distance, beginning at 22 pfennigs for express and 11 pfennigs for ordinary shipments. The "ordinary" carload rate for all freight not claiming a special rate is 6 pfennigs per kilometer. But the benefit of "special" tariffs is accorded to ten-ton carload shipments of a large number of articles. Special Tariff I., relating principally to manufactured products, is 4.5 pfennigs; Special Tariff II., is 3.5 pfennigs; Special Tariff III., relating principally to raw materials, is 2.6 pfennigs for distances up to 100 kilometers, 2.2 pfennigs thereafter. In the case where the shipment is more than 5 but less than 10 tons, these special tariffs are increased. These are all "normal" rates. The most important "exceptional" tariff is that for raw materials (e.g., coal and iron ore): 2.2 pfennigs for distances up to 350 kilometers, above that 1.4 pfennigs.

[b] A large number of the exceptional tariffs are descended from the days of private railways, which were bought up 1879-80 in Prussia. An official Prussian document details reasons for which exceptional tariffs are retained or introduced: "the exceptional tariffs were in large part taken over from the private railways in order to protect large interests dependent on them, or have been occasioned by competition of other railways or waterways. Moreover, the economic interests of the Empire, of the state or of particular regions, have been the occasion of granting exceptional tariffs. This was done particularly:

has become the exception. We are particularly interested in one class of exceptional tariffs, the import and export rates, for they affect the German seaports.

Most of the export rates apply to all exports from Germany, whether by sea or land; but, in case the shipment is destined for a seaport, the export rate is usually allowed only when the port is a German one. This condition is attached to L. C. L. shipments for export by sea. When sent over German ports,[a] they are accorded the ton-kilometric rate for full carload shipments of general merchandise, 6.7 pfennigs, instead of the normal rate of 22 to 11 pfennigs, varying with the distance. This reduction, and its restriction to German ports, is an important one for them, because so large a part of German exports consists of specifically valuable, not bulky goods, often sent in less than carload lots.

Similarly effective are the rates accorded German manufactures, particularly iron and steel, exported via German ports. Examples of these, compared with the normal, inland rates, are as follows:[b]

"1. To further industrial or agricultural production through cheapening their raw materials.

"2. To further the sale of native products in inland regions threatened by foreign competition and particularly to further the export of German products to foreign lands.

"3. To aid German commercial centers, particularly German seaports, against the competition of foreign seaports.

"4. To aid national means of communication, especially the state railways, against the competition of foreign railways and waterways."

(Der preussische Landeseisenbahnrat in den ersten 25 Jahren seiner Tätigkeit 1883-1908. Denkschrift des Ministers der Oeffentlichen Arbeiten, 1908.)

[a] Grotewold, page 21.
[b] Thackera, page 44.

WATERWAYS AND RAILWAYS

INLAND AND EXPORT RATES FOR MANUFACTURES.

(*Rates per ton for shipment in ten-ton carload lots.*)

From	To	Kilo-meters	Miles	Articles	Export Rate Marks $	Normal Marks $
Cologne–Hamburg		430	267.2	Copper goods, lead in blocks, pipes.	13.20 = 3.14	26.80 = 6.38
Cologne–Hamburg		430	267.2	Zinc in sheets.	13.30 = 3.17	20.40 = 4.86
Cologne–Hamburg		430	267.2	Cotton goods.	15.30 = 3.64	26.80 = 6.38
Cologne–Hamburg		430	267.2	Machinery, machine parts, iron wares.	10.60 = 2.52	20.40 = 4.86
Cologne–Hamburg		430	267.2	Iron plates, locomotives, etc.	5.60 = 1.33	16.10 = 3.83
Frankfort–Hamburg		532	330.6	Machines, iron wares.	12.90 = 3.07	25.20 = 6.00
Frankfort–Bremen		459	285.2	Machines, iron wares.	11.30 = 2.69	21.90 = 5.21
Frankfort–Lübeck		577	358.5	Machines, iron wares.	13.90 = 3.31	27.20 = 6.47
Frankfort–Hamburg		532	330.6	Iron products (beams, etc.).	7.00 = 1.67	19.80 = 4.71
Frankfort–Bremen		459	285.2	Iron products (beams, etc.).	6.10 = 1.45	17.30 = 4.12
Frankfort–Lübeck		577	358.5	Iron products (beams, etc.).	7.50 = 1.79	21.40 = 5.09
Nürnberg–Hamburg		635	394.6	Thuringian wares: toys, etc.	24.50 = 5.83	39.20 = 9.33
Nürnberg–Bremen		583	362.3	Thuringian wares: toys, etc.	22.90 = 5.45	36.40 = 8.66
Nürnberg–Lübeck		652	405.1	Thuringian wares: toys, etc.	25.30 = 6.02	40.50 = 9.64

An instance of the deflection of exports to Hamburg and Bremen is afforded by the growth of their receipts of manufactured iron from West Germany. West Ger-

THE PORT OF HAMBURG

many (the Prussian provinces of Rhineland, Westphalia, Hesse and Hesse-Nassau) is the chief seat of the German iron industry. It naturally exports via Rotterdam and Antwerp, reaching them most cheaply by transshipping from rail to barge in a Rhine harbor like Duisburg or Düsseldorf. It is by this route that West Germany sends most of its manufactures abroad: 798,760 tons in 1907.[a] However, the high normal rates are charged for hauling goods to Rhine river ports for transshipment. As a result, it is cheaper for many exported goods not originating within a short distance of the Rhine to take their way via Hamburg and Bremen, to which the export rates apply. In 1890 Hamburg and Bremen received 156,100 tons manufactured iron from western Germany.[b] In 1907 they received 502,800 tons.

German export rates extend in general only to the seaport; there is no prorating with the steamship line. A forwarder at the seaport must attend to the transfer of these exports from rail to ship and get a new ocean bill of lading. Two departures from this have been made: the prorating agreement with the German Levant Line in Hamburg and the Atlas Line in Bremen, the former and earlier made in 1890; and with the German East African Line, made in 1895. A through rate, absorbing the transfer charges at the seaport, is given from interior points in Germany to seacoast and even inland points in the Levant, for some foreign railways, such as the Roumanian and the Bulgarian state railways, join in the

[a] Bericht der Zentralkommission für die Rheinschiffahrt, 1907. This is the tonnage passing Lobith (Dutch border) downstream.

[b] Traffic districts 21-28 in Statistik der Güterbewegung auf deutschen Eisenbahnen. The 1890 figures are from Cords.

WATERWAYS AND RAILWAYS

agreement. The railway and the steamship companies each so reduces its rates for these shipments that in some cases a cheaper through rate may be obtained by shipping from inland than by delivering the merchandise to the steamer at Hamburg.[a] The result of these rates is to attract to Hamburg and Bremen shipments that would normally go via Italian and Austrian Mediterranean ports. Identical with this Levant prorating agreement with the two steamship lines first named is the agreement with the German East African Line.

The effect of the Levant prorating agreement is indicated by the great increase in the volume of Hamburg's exports to various countries of the Levant:[b]

HAMBURG'S EXPORTS TO THE LEVANT.

	1890 Marks	1899 Marks	1907 Marks
Turkey in Europe,	1,449,000	7,207,000	40,555,000
Bulgaria and Servia,		851,000	3,126,000
Russian ports on Black Sea,	1,671,000	8,236,000	11,749,000
Greece,	1,086,000	2,662,000	6,109,000
Roumania,	1,050,000	2,424,000	3,447,000
Asiatic Turkey,		1,987,000	10,073,000
	5,256,000	23,367,000	75,059,000

There is plenty of evidence concerning the stimulus which these prorating reductions give to the German export trade. Aftalion[c] states that from the interior of Germany shipments are made to Oriental ports costing

[a] Grotewold, page 34.
[b] Hamburgs Handel und Schiffahrt, 1907. Earlier figures reprinted in Aftalion, page 184.
[c] Aftalion, page 184.

THE PORT OF HAMBURG

four or five marks, which, if they were exported from an inland town in France, via Marseilles, would cost 15 to 20 francs. A British Royal Commission finds a similar advantage held by the German over the English exporter:[a]

"The differences in favor of German combined land and sea through rates of freight as against the British rates will be best appreciated by a perusal of some of the appendices . . . . but a few examples may be given here. The cost per ton of sending iron rails from Birmingham to Liverpool, which is 97 miles, or 156 kilometers, is 8s. 4d.; the cost per ton of sending iron rails from Oldenburg to Hamburg, which is 160 kilometers, is 3s. 4d. on the East African tariff and 3s. 2d. on the Levant tariff. Again, the cost of the carriage of machinery packed for export from Leicester to Glasgow, 313 miles, or 504 kilometers, is 36s. 4d. a ton; from Wronke, East Prussia,[b] to Hamburg, also 504 kilometers, it is 10s. 10d. a ton on the East African tariff and 7s. 10d. a ton on the Levant tariff. The carriage of hardware from Birmingham to London, 111 miles, or 179 kilometers, is 21s. 8d. a ton; from Flensburg, Schleswig-Holstein, to Hamburg, also 179 kilometers, it is 4s. 2d. a ton on the East African tariff and varies according to the kind of hardware between 4s. 1d. and 4s. 9d. on the Levant tariff. The rate for export bales of cotton from Manchester to Bristol, 175 miles, or 282 kilometers, is 22s. 4d. a ton; from Berlin to Hamburg, 279 kilometers, it is 8s. 7d. on the East

---

[a] 1902 Report of the Royal Commission on Steamship Subsidies, reprinted in 1903 Report of the United States Commissioner of Navigation, page 270.

[b] Of course no machinery is exported from East Prussia, an agricultural region pure and simple; yet the force of the example is not weakened.

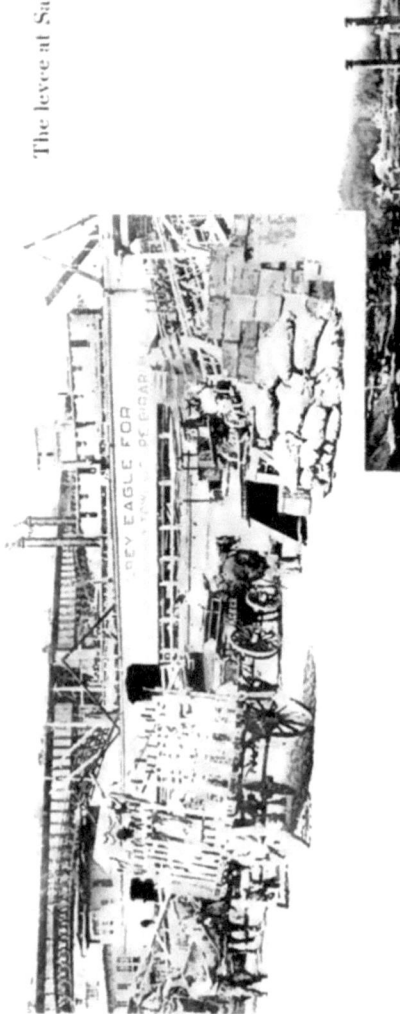

The levee at Saint Louis.

A cotton landing on a tributary of the Mississippi.

Mississippi river scenes.

WATERWAYS AND RAILWAYS

African tariff and 4s. 1d. on the Levant tariff. Plows and cement can be shipped to Delagoa Bay rather more cheaply by British than by German routes, but the German through tariff to Delagoa Bay is cheaper, and in most cases far cheaper, for bicycles, bottles, brush ware, carpets, galvanized iron, heavy iron work, glass furniture, hollow ware, iron pipes, nails and pianos. The practical effect of these cheap through rates upon British trade will be considered in a subsequent paragraph; suffice it to say, at present, that the reduced rates have been and are fixed in accordance with the experience gained in Germany as regards the working cost per train mile over long distances, and that the primary object is the building up, promoting or increasing of German export trade to the countries in question, and the enabling it to compete successfully with the trade of other foreign states to those countries. From the German point of view the policy of all these direct and indirect subsidies has been thought a good investment and worth the money spent, but an element of success is the energy and enterprise of the recipient."

Import rates are given for raw materials and foodstuffs not grown in Germany and are given exclusively for importations via German seaports. They are directly designed to protect Hamburg and Bremen against the competition of the Dutch and Belgian ports in western Germany. These ports are 125 to 150 miles nearer most points in Rhineland and Westphalia than Hamburg and, moreover, have to their advantage the use of that magnificent waterway, the Rhine, to communicate with Germany. The weapon of the railways is of course the application of high normal rates for distributing merchandise

THE PORT OF HAMBURG

transshipped from barge to rail in the Rhine river ports, as against low import rates direct from Hamburg and Bremen. Petroleum (in carloads) is sent inland from Hamburg and Bremen at the rate of 2.2 pfennigs per ton-kilometer, the normal rate is 4.5 pfennigs; cotton has an import rate of 2.2 pfennigs, the normal rate is 4.5 pfennigs; rice has an import rate of 2.7 pfennigs, the normal rate is 4.5 pfennigs; coffee, tobacco and raisins have an import rate of 3.5 pfennigs instead of 6 pfennigs.

These rates have been variously successful in holding the trade of West and South Germany and Switzerland[a] for German seaports. In petroleum they have failed. The invention of the tank barge gave a means of transportation to Rhine ports so cheap that they could pay high distributive rates and yet ship to inland points more cheaply than Hamburg and Bremen can.[b] In 1907 Hamburg and Bremen sent by rail to South and West Germany and Switzerland 49,250 tons of petroleum.[c] Up the Rhine past the Dutch-German border in the same year came 313,300 tons petroleum,[d] with the same destination. The classic example of the working of import tariffs restricted to German ports is the case of imported cotton. In 1907 Bremen sent to the territory just mentioned 274,400 tons of raw cotton; on the Rhine were imported only 30,000 tons. Up the Rhine into Germany came 42,000 tons coffee; Hamburg alone sent to the Rhine

[a] Traffic districts 21-37 and 56 of the Statistik der Güterbewegung.
[b] There is, to be sure, a coöperation between the South German railways and the upper Rhine ports (Mannheim, etc.), such as we observe the upper Elbe ports enjoying.
[c] Statistik der Güterbewegung auf deutschen Eisenbahnen, 1907.
[d] Jahresbericht der Zentralkommission für die Rheinschiffahrt, 1907.

WATERWAYS AND RAILWAYS

territory 30,350 tons. Later, in Chapter VIII., we shall have more detailed examples of the distances to which Hamburg's inland rail shipments penetrate.

Transit tariffs exist, designed to attract and hold for the German seaports and the German railways oversea imports destined for points beyond the German border. For instance, cotton, instead of the normal rate of 4.5 pfennigs, pays 3 pfennigs from German ports to Bohemia, 2.2 pfennigs to Poland, 1.75 pfennigs to Russia. Rice has a reduction from the same normal tariff to 2.69 pfennigs to Austria, 2.61 pfennigs to Poland, 2.51 pfennigs to Galicia, etc. The effect of these transit tariffs is to extend to foreign countries the hinterland that is served by German ports.

Hamburg, then, has in the Elbe and its connecting waterways a splendid means of communication with all eastern and central Germany. Primarily for this reason, Hamburg predominates in the foreign trade of central Germany and draws heavily on the eastern part of the Empire. The greatest advantage is taken of the opportunity to use the waterways. Huge sums have been spent on their betterment, the floating stock of the river modernized, transportation organized under large companies, and—most important of all—modern terminal facilities provided for the barges both in Hamburg and in the river ports. The tariffs of the Prussian railways, mildly hostile to coöperation with the Elbe, are counterbalanced by the coöperation of the Saxon and Austrian railways.

South and West Germany, with which Hamburg has no waterway connection, are drawn to Hamburg by the friendly tariffs of the Prussian railways, which limit the application of their low export and import rates to ship-

THE PORT OF HAMBURG

ments via German seaports. Especially needed—and effective—were these tariffs to prevent the Rhine from carrying off to foreign seaports the trade of the richest part of the Empire. The working of the tariffs that we have discussed shows how, in the hands of the state, the railways are a valuable instrument to use in furthering the aims which the nation sets before itself.

CHAPTER VII

SHIPPING AND COMMERCE IN HAMBURG
1907

ALONG with the growth of Germany's foreign trade, detailed in Chapter I., went a growth in the frequency with which her seaports were sought out by vessels engaged in that trade. The increase in the total movement of shipping in German ports is indicated by the following table:[a]

TONNAGE OF SHIPS ENTERING AND LEAVING GERMAN PORTS.

	Steamers		Sailers		Total	
	Number	1,000 Net Register Tons	Number	1,000 Net Register Tons	Number	1,000 Net Register Tons
1873	17,089	6,439	77,598	5,903	94,687	12,342
1895	65,970	26,124	67,860	4,345	133,830	30,469
1907	140,005	50,398	68,771	5,739	208,776	56,129

In thirty-five years the movement of shipping in German ports has more than quadrupled, the tonnage of steamers has increased eightfold. The tendency is more and more for the German flag to predominate.

FLAGS OF SHIPPING ENTERING AND LEAVING GERMAN PORTS.[b]

	I Vessels under German Flag		II Vessels under Foreign Flag		III Vessels under English Flag (included in II.)	
	Number	1,000 Net Register Tons	Number	1,000 Net Register Tons	Number	1,000 Net Register Tons
1873	60,342	5,964	34,345	6,378	9,361	3,384
1895	97,375	15,938	36,455	14,451	11,453	9,555
1907	159,673	31,699	49,103	24,430	12,004	12,343

[a] Statistische Jahrbücher für das deutsche Reich. The present growth of the German merchant marine is phenomenal. It increased in the period 1900-10 from 2.9 to 4.9 million gross register tons, the steam tonnage from 2.1 to 4.2 million tons.

[b] Ibid.

THE PORT OF HAMBURG

In 1873 more goods were shipped under foreign flags than under the German; the proportion of shipping carrying the German flag was 46 per cent. In 1907 the situation was reversed; 56 per cent of the shipping entering and leaving German ports went under the national flag.[a]

The increase in the movement of shipping has been general in German ports, but they have not all partaken of this increase in the same degree. The development of Germany into an "industrial" state, increasingly dependent on foreign trade, has occurred primarily in central and West Germany, so that ports on the North Sea were in the best position to have the advantage of it.

NET REGISTER TONNAGE OF VESSELS ARRIVING
IN GERMAN PORTS.[b]

	1871	1907
North Sea		
Hamburg,	1,887,500	12,040,400
Bremen,	886,000	3,570,500
Baltic Sea		
Stettin,	416,200	2,160,400
Dantzig,	447,300	887,100
Kiel,	203,400	623,200
Lübeck,	220,800	736,700
Königsberg,	278,300	706,300

Assuming the tonnage of 1871 in each case to be 100, the index numbers showing the increase in tonnage 1871-1907 are as follows:

---

[a] Not since 1886 have American vessels in any one year carried over 15 per cent of our exports and imports. In the years 1908-09-10, American ships carried 9.8, 9.5 and 8.7 per cent, respectively. (1910 Report of Commissioner of Navigation, page 172.)

[b] Statistische Jahrbücher für das deutsche Reich, and 1907 Bericht der Aeltesten der Kaufmannschaft, Stettin.

SHIPPING AND COMMERCE

INDEX NUMBERS OF INCREASE IN SHIPPING AT
GERMAN PORTS.

	1871	1907
Hamburg,	100	635
Bremen,	100	412
Stettin,	100	519
Dantzig,	100	199
Kiel,	100	306
Lübeck,	100	335
Königsberg,	100	254

In spite of their smaller initial tonnage, the ports of the Baltic have increased relatively more slowly than those of the North Sea. Stettin is of great importance in the Russian and the Scandinavian trade but it still has no oversea connections. It is only about half as far as Hamburg is from Berlin—a city of two million—and from the Silesian industrial district. It is possible that, as the river Oder is improved and the Finow Canal from Berlin to Stettin is replaced by one of modern dimensions, Stettin will become a third great German port. As it is, Hamburg and Bremen, the North Sea ports, have increased more rapidly, both absolutely and relatively, than the Baltic ports.

Particularly noticeable for absolute and relative increase has been the port of Hamburg. Though in 1871 it started with a larger tonnage than its rivals, it has grown more rapidly than any of them—more rapidly than all of them together. This is due partly to its seaward position on the trade-ward side of the Jutland peninsula, but primarily to its incomparable position with reference to the common German hinterland, for which all the ports compete. The excellence of Hamburg's water and rail connections with this hinterland has already

THE PORT OF HAMBURG

been detailed. As a result of this, Hamburg in 1907 had in its basins a movement of vessels totaling twenty-four million tons, 40 per cent of the total shipping in German ports.

Hamburg is now, after London, the first port of Europe. The following table shows the tonnage of ships entering the principal European ports in 1907:

TONNAGE OF SEASHIPS ENTERING EUROPEAN PORTS IN 1907.[a]

Port	Net Register Tons
London,	11,160,000
Liverpool,	8,167,000
Hamburg,	12,040,000
Antwerp,	11,200,000
Rotterdam,	10,107,000

In this table, vessels in the coasting trade of the continental ports are included; in the case of the English ports, they are not.[b] England is a small island and coasting vessels there do a large part of the collecting and distributing which on the continent is done by inland waterways and railways. Antwerp's method of calculating net register tonnage makes its total 15 per cent larger than it would be if Antwerp followed the method of other ports. Omitting Hamburg's coasting trade—1,152,000 tons—from the total, it had 10,898,000 tons entering in 1907, engaged in the foreign trade. This puts Hamburg second to London alone.

[a] Compiled from the publication "Office de statistique universelle d'anvers. Annuaire 1908."

[b] If vessels in the English coasting trade were included, London, with a total of 17,292,000 tons, would take first place among European ports, Liverpool third place, after Hamburg.

SHIPPING AND COMMERCE

In 1907 there entered and left Hamburg vessels aggregating 24,143,700 net register tons. Arranged by arrivals and departures this total may be divided as follows:[a]

MOVEMENT OF SHIPPING IN HAMBURG, 1907.

	Number of Vessels	Net Register Tons
Arrivals,	16,473	12,040,400
Departures,	16,507	12,103,200

Of this number of departures, 4,572 ships of 4,078,200 tons left in ballast, 28 per cent of the tonnage of the ships that sailed. But a very large proportion of these were tramp vessels leaving for England to get a cargo of coal,[b] or colliers regularly engaged in carrying English coal to Hamburg (four and one half million tons in 1907).

[a] Hamburgs Handel und Schiffahrt.
[b] It is well known what a treasure the English coal exports are to the English merchant marine, in furnishing it with outbound freight for vessels that have brought bulky raw materials and foodstuffs to England. In 1907, 87 per cent of the volume of England's exports consisted of coal. The following table shows the production and exportation of coal in Germany and England, respectively, in 1907:

Production and Exportation of German and English Coal, 1907.

	Production	Exports
Germany,	143,000,000 Tons	20,000,000 Tons
England,	253,000,000 Tons	75,000,000 Tons

(Charles et de Rousiers, page 122.)

Germany's coal production has hardly been able to keep pace with the national consumption. Moreover, part of Germany's exports go by rail, all of England's by sea. Hamburg is not so situated that it can partake of the German coal export—it is too far away from all the coal mining districts. Hamburg is more interested in importing English than exporting German coal.

THE PORT OF HAMBURG

Three thousand and eight ships of 3,841,000 tons left in ballast with England as their destination.

The increase in the tonnage of vessels entering Hamburg since 1850 is shown in the following table:

VESSELS ARRIVING IN HAMBURG.

	Number	Net Register Tons
1851-60, average	4,649	756,100
1861-70, average	5,092	1,260,700
1871-80, average	5,502	2,206,300
1881-90, average	7,015	3,870,000
1899	13,312	7,765,900
1907	16,473	12,040,400

This progress has been accomplished under the German flag. In 1871 Hamburg was still tributary to the British merchant marine, but an emancipation has since taken place. In 1896, for the first time, the German flag predominated in the port. From that year the progress has been steady: in 1907 the German flag floated over 53 per cent of the tonnage of vessels arrived, the English flag over 35 per cent; in the average of the years 1871-80 these percentages were precisely reversed.

PERCENTAGES OF HAMBURG, FOREIGN AND ENGLISH FLAGS IN TONNAGE OF VESSELS ENTERING HAMBURG.[a]

	English Flag Net Register Tons	German Flag Net Register Tons	Hamburg Flag in particular Net Register Tons	Percentages		
				English %	German %	Hamburg %
1871-80, average	1,173,400	747,000	577,400	53	35	26
1907	4,209,500	6,395,500	5,104,600	35	53	42

[a] Hamburgs Handel und Schiffahrt.

SHIPPING AND COMMERCE

The part played by other nationalities in Hamburg's shipping is insignificant, the Scandinavian flags being those most frequently seen.

As is evident from a glance at the percentages in the table above, the increase in the percentage of vessels carrying the German flag was due to those vessels carrying the Hamburg flag. The growth of Hamburg's merchant marine has been as follows:[a]

STEAM VESSELS.

	Number	Gross Tons	Net Tons	Crews
1885	187		186,500	5,495
1900	435	993,709	624,200	13,888
1910	704	2,133,300	1,305,000	28,336

STEAM AND SAILING VESSELS.

	Number	Gross Tons	Net Tons	Crews
1885	480		319,471	8,800
1900	846	1,229,500	846,634	17,700
1910	1,325	2,417,615	1,586,900	33,000

The increase has been almost entirely in the tonnage of steam vessels. The Hamburg fleet has come to represent 50 per cent of the tonnage of the German merchant marine, while in 1885 the corresponding percentage was eighteen.

We have already seen (page 100) how extensive are Hamburg's connections, by means of regular lines, with the oversea world. The following is a list of the regular lines operating out of the port of Hamburg in the year 1907:[b]

[a] Handbuch der deutschen Handelsmarine.
[b] Hamburgs Handel und Schiffahrt.

THE PORT OF HAMBURG

REGULAR LINES SERVING HAMBURG.

To	Total	Lines under German Flag
German ports,	14	14
England,	33	9
Russia,	7	5
Sweden,	8	3
Norway,	4	0
Denmark,	4	1
Holland,	4	1
Belgium,	2	2
France,	2	0
Mediterranean Europe,	7	5
North America,	5	4
Central America,	3	2
South America,	12	11
Africa,	9	7
Asia, Australia,	14	12

SUMMARY.

	Lines	Steamers	Voyages
European service,	84	556	7,032
(German,	45	290	3,746)
Extra-European service,	44	562	1,180
(German,	36	481	998)
Total,	128	1,118	8,212
(German,	81	771	4,744)

But of course a port's shipping is not confined to ships that ply in regular lines. A more complete picture of Hamburg's oversea shipping is furnished by an analysis— by countries—of the total movement of vessels in Hamburg in 1907. Combined with similar figures for 1899, they show the line of development that Hamburg's shipping is taking.

In the first group, European countries, notice should be taken of the considerable increase in the tonnage of ships departing for Scandinavia as well as of the increase

SHIPPING AND COMMERCE

ORIGIN AND DESTINATION OF SHIPPING IN HAMBURG, 1907 (1899).[a]

1000 Register Tons.

Country	Arrived 1907	(1899)	Sailed 1907	(1899)
Scandinavia,	560	(403)	635	(478)
North Russia,	132	(56)	215	(124)
Spain-Portugal,	218	(108)	104	(85)
Italy,	150	(93)	53	(44)
Levant,	732	(330)	120	(51)
U. S. A.,	1,833	(1,508)	1,543	(1,271)
U. S. A. (east coast),	1,803	(1,508)	1,486	(1,238)
Canada,	9	(45)	11	(61)
West Indies, Central America,	263	(177)	356	(163)
South America (east coast),	915	(417)	651	(395)
South America (west coast),	388	(383)	297	(215)
Africa except Mediterranean,	558	(258)	792	(291)
British India and East Asia,	744	(380)	530	(316)
Australia,	147	(78)	160	(115)

in the tonnage of vessels arriving from the Levant.[b] Roumania and Russia are sending their grain, as America is no longer able to meet the growing demand in Germany for foreign grain. In the second group, beside a steady increase in the tonnage outbound to the United States and inbound from there, the tonnage engaged in the trade with South America—east coast—is striking. That means grain and wool from Argentine, coffee and rubber from Brazil, and German manufactures sent to pay for them. In the South American trade the increase in inbound tonnage, both relatively and absolutely, is the largest in the whole list. In the fourth group we are struck by the growth of arrivals from British India and Asia, and the increase of tonnage sailing for Africa. The

[a] Hamburgs Handel und Schiffahrt.
[b] By the Levant is meant countries on the Black Sea, the Balkan countries and Greece, as well as Asia Minor.

THE PORT OF HAMBURG

latter is an indication of Germany's intention to play a leading rôle in the commerce of the newest continent in international trade.

While the net register tonnage—the carrying power—of vessels entering or leaving port is only an indication of the intention to carry goods, yet ships do not throng a port without having cargoes or prospects of getting them. The activity of shipping in Hamburg has been an intimation of the enormous commerce that we should find developed in the port. In 1907 this commerce attained the following dimensions:

HAMBURG'S COMMERCE, 1907.[a]

	Weight	Value
Imports,	14,670,200 Tons	3,577,314,700 Marks
Exports,	6,142,200 Tons	2,802,218,100 Marks
Total,	20,812,400 Tons	6,379,532,800 Marks

The disparity of eight and one half million tons between the weight of imports and exports is what makes so many of the tramp vessels leave for England in ballast to get an outbound cargo of coal. In 1907 ships brought to Hamburg four and one half million tons of coal, two and one half million tons of grain, 600,000 tons of feed, 500,000 tons of saltpetre, 450,000 tons of mineral oils, 275,000 tons each of flaxseed, pig iron, rice and copra. Hamburg has few bulky goods to fill the ships that have brought these imports. In 1907 she exported 975,000 tons of sugar, 550,000 tons of potash, 425,000 tons of grain, 300,000 tons of cement and 250,000 tons of manufactured iron: no other article of export exceeded 200,000

[a] Hamburgs Handel und Schiffahrt. These figures include the coasting trade, a small part of the total trade.

Loading a trainload of cable direct from cars into steamer, at a pier of the Hamburg-American Line, at Kuhwärder.

SHIPPING AND COMMERCE

tons. We remember, however, that the return freight of the liners in the transatlantic service is emigrants, who on the outbound trips fill the upper cargo decks. In 1907 Hamburg had 190,000 emigrants to export.

The value of Hamburg's commerce is 2.7 billion dollars, larger than the total foreign commerce of many European states, such as Switzerland and Spain, and about equal to that of Italy. Sixty per cent[a] of the maritime commerce of Germany, 40 per cent of its entire foreign commerce, is carried on through Hamburg. A single port, a single point on the coast of the North Sea attracts to itself 40 per cent of the commerce that passes through all the ports of a coast line that measures 1,530 miles and over terrestrial borders 5,230 miles long. The tables of Hamburg's commerce in the years 1871-1907 bear record to a remarkable growth.

GROWTH OF HAMBURG'S COMMERCE.[b]

	Imports		Exports		Total	
	Weight 1,000 Tons	Value Million Marks	Weight 1,000 Tons	Value Million Marks	Weight 1,000 Tons	Value Million Marks
1871-80, average	2,102	874	968	597	3,170	1,471
1881-90, average	3,496	1,046	2,000	981	5,496	2,027
1891-05, average	5,756	1,559	2,693	1,267	8,449	2,826
1899 . .	9,178	1,984	4,155	1,606	13,353	3,590
1907 . .	14,670	3,577	6,142	2,802	20,812	6,379

Save that the progress is more striking in Hamburg than in the whole Empire, there is a plain parallelism

[a] In 1905 the official publication "Die Entwicklung der deutschen Seeinteressen" estimated that 70 per cent of Germany's foreign commerce was sea-borne. Seventy per cent of Germany's foreign commerce in 1907 (15,586 million marks) is 10,910 million marks. Hamburg's quota of this, as is seen from the table above, is 60 per cent.

[b] Hamburgs Handel und Schiffahrt.

THE PORT OF HAMBURG

between the expansion of Germany's foreign commerce and of Hamburg's. This parallelism, followed in detail, shows that Hamburg's prosperity is a result of the economic evolution of Germany. The characteristic of Germany's transformation into an industrial state is, as we have seen, the increase in its imports of raw materials and foodstuffs, the increase in its exports of manufactured products. The same characteristic distinguishes the commerce of Hamburg. Its leading imports in 1907, in point of value, were as follows:

HAMBURG'S LEADING IMPORTS, 1907.[a]

Article	Value Million Marks
Grain,	298.2
Wool,	205.6
Coffee,	204.9
Cotton,	160.4
Cow hides,	112.7
Saltpetre,	111.8
Rubber,	111.3

The progress of Hamburg's importation since 1870, according to the main groups of imports, has been as follows:

HAMBURG'S IMPORTS, 1871-1907.[b]

	Foodstuffs		Raw Materials		Manufactures	
	1,000 Tons	Million Marks	1,000 Tons	Million Marks	1,000 Tons	Million Marks
1870	135	93	1,109	299	63	121
1880	552	314	1,848	446	97	134
1890	1,228	469	3,655	774	123	133
1899	2,744	631	6,211	1,148	223	204
1907	4,246	1,082	10,082	2,166	342	329

[a] Hamburgs Handel und Schiffahrt.
[b] Hamburgs Handel und Schiffahrt. Earlier figures reprinted in Aftalion, page 574.

SHIPPING AND COMMERCE

Imports of foodstuffs have increased in volume more than thirty-fold, as compared with 1870; in value they have increased eleven-fold in spite of the great drop in prices that has occurred. The growth in the imports of raw materials has been still greater absolutely, though smaller relatively: these imports increased in volume nearly tenfold, in value over seven-fold. The progress in the importation of finished products is slower; their value has not trebled in thirty-seven years.

In 1907 foodstuffs constituted 30 per cent of the total value of imports in Hamburg, raw materials 60 per cent and manufactures only 10 per cent. We see mirrored here the growing need for foodstuffs of a population no longer capable of supporting itself and the growing demand for raw materials to supply its industries. It is the industrialization of Germany that is responsible for the nature of Hamburg's increasingly large imports.

Similar conclusions force themselves upon us when we examine the statistics of Hamburg's exports. The principal articles of export in 1907 were:

HAMBURG'S LEADING EXPORTS, 1907.[a]

Article	Value Million Marks
Sugar,	228.1
Manufactured Iron,	138.2
Chemicals and drugs,[b]	125. (circa)
Machines,	120.3
Cotton goods,	105.9
Woolen and half-woolen goods,	66.5
Coffee,	81.4
Paper,	60.
Hosiery,	55.4

[a] Hamburgs Handel und Schiffahrt.
[b] This is one of the most promising of Germany's exporting industries. The export of artificial indigo, for instance, increased from

THE PORT OF HAMBURG

The progress of Hamburg's exportation since 1870, according to the main groups of exports, has been as follows:

HAMBURG'S EXPORTS, 1871-1907.[a]

	Foodstuffs		Raw Materials		Manufactures	
	1,000 Tons	Million Marks	1,000 Tons	Million Marks	1,000 Tons	Million Marks
1880	919	374	479	183	124	248
1890	1,359	506	825	271	328	483
1899	2,001	544	1,672	485	482	576
1907	2,426	719	2,798	938	918	1,145

In contrast to the import tables, manufactures here occupy the first place among the groups, ahead of raw materials and foodstuffs. The value of the latter is still high, largely because nearly all Germany's sugar export, and a large part of Austria's, passes through Hamburg—to the value of 228 million marks in 1907. Yet the exports of foodstuffs have not trebled in volume since 1880; in value they have increased only 92 per cent. The value of manufactured goods exported has grown 363 per cent since 1880.[b] These exports of finished products are to pay for the imported raw materials and foodstuffs. In its exports as well as its imports Hamburg's commercial development follows the industrial development of Germany.

We may, therefore, expect to find Hamburg importing primarily from producers of foodstuffs and raw materials

1905-07 from three and one half million to thirteen and one half million marks.

[a] Hamburgs Handel und Schiffahrt. Earlier figures are reprinted in Aftalion, page 575.

[b] The increase in the exportation of raw materials is largely due to Hamburg's transshipment trade. Instances will be given later.

SHIPPING AND COMMERCE

and exporting to free trade countries like England, or agricultural states, not yet industrially developed, like Brazil. Such is the rule, Great Britain being the exception among the leading sources of imports, the United States among the destinations of exports.

HAMBURG'S IMPORTS.[a]

Leading Countries	Million Marks 1899	1907
South America,	327	683
England,	421	635
United States,	446	589
British India,	147	350
Russia,	71	182

HAMBURG'S EXPORTS.

Leading Countries	Million Marks 1899	1907
England,	434	514
South America,	174	400
United States,	166	347
Scandinavian States,	163	266
Russia,	64	142
China, Japan,	61	116

In the following analysis the bracketed figures are those applying to the year 1899, the others to the year 1907. South America sends from Argentine 66 (50) million marks of wool, 50 (15) million marks of cereals and 47 million marks of flaxseed; from Brazil 133 (60) million marks of coffee and 24 million marks of rubber; from Chili 111 million marks of saltpetre. Great Britain sends Hamburg textiles, machines and coal. Hamburg imported 184 (120) million marks of foodstuffs from the United States, among them 70 (100) million marks of cereals and 53 million marks of lard. Moreover, the United States sends 328 (315) million marks of raw materials[b] and only 75 million marks of manufactured goods—45 million marks machines. From India Hamburg gets cotton and jute to the value of 87 and 73 mil-

[a] Hamburgs Handel und Schiffahrt. 1899 figures reprinted by Aftalion, page 576.

[b] Seventy-six million marks of copper, 51 million marks of raw cotton and lint, 28 million marks of petroleum.

lion marks, respectively. From Russia come cereals to the value of 112 (30) million marks, also petroleum.

In the export table Hamburg sends England 175 (156) million marks of sugar, 137 million marks of raw materials and 153 (130) million marks of finished products. Almost the entire exportation to South America consists of manufactured goods. To Scandinavian states, besides exporting manufactured products, Hamburg sends a large number of "colonial" wares, wines, nitrates and other foreign goods which are transshipped in Hamburg's Free Port. In this re-exportation trade appears the value of the Free Port; it represents, moreover, the modern remains of Hanseatic days when Hamburg was an entrepôt for the countries of the north.

As Hamburg's commercial statistics deal only with the "general" commerce and do not distinguish re-exportations from "special" exports—of goods originating in Germany—it is difficult to ascertain the exact amount of Hamburg's transshipment or re-exportation trade. However, we can be certain that all exports belong to this re-exportation class when they consist of goods not produced in Germany. The more important among the articles of this transshipment trade are given in the table below; the figures for export, in combination with those giving the total imports of these goods, indicate how large a part transshipment has in disposing of such imports.

HAMBURG'S EXPORTS OF NON-EUROPEAN PRODUCTS, 1907.[a]

	Million Marks	
	Imported	Exported
Coffee,	205.	81.5
Wool,	205.6	49.
Vegetable oils,	42.4	47.

[a] Hamburgs Handel und Schiffahrt.

SHIPPING AND COMMERCE

	Million Marks	
	Imported	Exported
Rubber,	117.4	39.5
Cotton,	160.4	38.
Cacao,	72.5	35.
Tobacco,	56.4	32.8
Rice,	47.3	31.2
Lead and copper,	127.	28.5
Saltpetre,	112.	28.
Petroleum and lubricating oil,	55.5	22.
Copra,	94.5	21.6
Jute,	75.7	20.
Maize,	68.7	18.
Lard,	53.8	16.4
Fish,	27.7	16.
Wine,	25.6	10.

These are extra-European products from lands across the sea with which, in many cases, the countries of north-eastern Europe have little trade. Hamburg's exports to these distant lands bring their products back to the Elbe as return freight. In Hamburg, markets have developed for the colonial wares; here, also, the Free Port offers excellent facilities for storing, manipulating and re-shipping the goods. In the case of a considerable number of the articles, half or more of the quantity imported is re-exported; that is, foreign countries are better buyers than Germany itself. This is the case with rice, tobacco, cacao and vegetable oils. Of the latter more is exported than imported, due to the activity of the oil mills in the Free Port, which transform into oil the great quantities of oil seed brought to port, to send either inland or to foreign countries.

Yet this transshipment trade, this re-exportation, constitutes only a fraction of the total exports of the port. A general view of the disposal of imports—whether sent inland, consumed or re-exported—may be had from a

THE PORT OF HAMBURG

comparison between the value of Hamburg's imports in 1907, by main divisions, and the value of goods sent inland, similarly grouped. The difference represents the value of goods consumed by the population and industries of Hamburg or re-exported by sea. This table for 1907 is as follows:

HAMBURG'S IMPORTS AND DISPOSAL OF THEM, 1907.[a]

	Million Marks			
	Food-stuffs	Raw Materials	Finished Products	Total
Imported,	1,082	2,166	329	3,577
Sent inland,	612	1,493	325	2,380
Consumed in Hamburg or re-exported,	470	723	4	1,197

Probably half of this surplus is consumed by a manufacturing city of 900,000 souls, the other half re-exported. Among the re-exports in the first group are, of course,

[a] Hamburgs Handel und Schiffahrt. Similar figures indicating the part Hamburg's hinterland plays in the port's exports are the following:

Hamburg's Exports and the Source of Them, 1907.

	Million Marks			
	Food-stuffs	Raw Materials	Finished Products	Total
Exported,	719	930	1,145	2,802
Arrived (rail or river),	483	573	1,306	2,262
Re-exportation,	236	365		540

However, this does not represent the entire re-exportation, for many of the goods brought to Hamburg by river and rail were for local consumption. It is interesting to compare the total value of all goods arriving in Hamburg by sea, river and rail with the value of all goods similarly sent from the city. We discover that in 1907 the city brought from without for consumption 234 million marks of foodstuffs, 358 million marks of raw materials and 155 million marks of manufactures.

SHIPPING AND COMMERCE

coffee, tobacco, rice and cacao; in the second group, wool, cotton, vegetable oils and saltpetre. In the third group, manufactured products, there is practically no surplus.

In 1903 Cords[a] estimated that one million tons, or 19 per cent of the volume of exported goods, consists of re-exports. A similar estimate for 1907 would make it appear that 1,350,000 tons of goods were re-exported, 22 per cent of the total exports. However, Cords' method of estimating is a rough one and no conclusion should be drawn from the figures given, as to the tendency of the re-exportation trade to increase relatively to the total volume of exports. Charles and de Rousiers[b] estimated in 1906 that the value of re-exports was over 16 per cent of the whole value of exports. In general, it can be said that about one fifth of Hamburg's exports in volume—and probably in value—are re-exportation; that, in contrast with conditions a century ago, Hamburg has become four times more interested in exporting for her hinterland than in serving as an entrepôt for foreign nations.

We have seen, then, that Hamburg is today the first port of Europe, with the single exception of London. In all German seaports there has been, since the Empire was formed, a large growth of shipping, in which the German flag has come to preponderate. This increase is seen to be largest in the ports of the North Sea, Hamburg and Bremen, particularly Hamburg, which in 1907 had 40 per cent of the total movement of vessels in German harbors. In Hamburg, as in Germany in general, the German flag dominates. Hamburg's merchant fleet and her line connections with foreign lands show a phenomenal increase,

[a] Die Bedeutung der Binnenschiffahrt, page 125.
[b] Le port de Hambourg, page 136.

THE PORT OF HAMBURG

as do the statistics of her commerce. The principal articles of the Elbe port's commerce are, among imports, raw materials and foodstuffs; among exports, manufactured goods—what we should expect of an "industrial" state. Re-exportation constitutes only about one fifth, in value and bulk, of the export trade. Hamburg is importing and exporting for its hinterland.

CHAPTER VIII

HAMBURG'S COMMERCE WITH ITS HINTERLAND
1907

IN FOLLOWING the distribution and collection of the articles of Hamburg's seaward commerce, by regions or by waterway and railway respectively, we are dependent on three statistical sources: the Statistics of Traffic on German Railways, the Imperial Statistics of Internal Navigation, and Hamburg's Trade and Shipping.[a]

The railway statistics are excellent. Germany is divided into thirty-seven traffic districts and the freight movement is recorded between each of these districts and all others, in 101 different articles, with subdivisions. Moreover, a similar record is kept of the traffic between each of these districts and various adjacent foreign countries or provinces. The German districts correspond, in general, with the Prussian provinces and the other German states; one district is represented by the Elbe seaports. Practically this means Hamburg. Harburg and Altona are the more important of the other "Elbe seaports," but their proportion of the tonnage in question is small; moreover, they are dependents of Hamburg. Thus we know to what portion of Germany, Austria, Russia and Switzerland Hamburg sends goods by rail, what goods, and in what amounts.

In Hamburg's Trade and Shipping there is given in

[a] Statistik der Güterbewegung auf deutschen Eisenbahnen, Die Binnenschiffahrt (Reichsstatistik), and Hamburgs Handel und Schiffahrt. The first is published by the Prussian Ministry of Public Works, the second by the Imperial Statistical Office, the third by the state of Hamburg.

THE PORT OF HAMBURG

detail the amount and value of Hamburg's shipments inland, by Elbe and by rail. Hamburg's receipts from inland are noted in similar detail. As regards rail shipments, there is no indication of their destinations; this must be gathered from the Railway Statistics. The total tonnage of shipments by the Elbe is distinguished as destined for the Oder, Havel, Prussian Elbe, Anhalt Elbe, Saale, Saxon and Bohemian Elbe and Elbe-Trave Canal (a canal from the Elbe to Lübeck). The total river shipments to the chief river ports are also recorded: Tangermünde (below Magdeburg), Berlin, Magdeburg, Schönebeck (just above Magdeburg), Dresden, Aussig, Schönpriesen and Laube-Tetschen—the last three on the Bohemian Elbe.

In the Imperial Statistics for Internal Navigation detailed information is given in regard to the receipts and shipments of a large number of German river ports. These statistics may in many cases be advantageously combined with those of Hamburg's Trade and Shipping; for instance, in order to discover the destinations of cotton and petroleum which Hamburg sends inland by water. Trade and Shipping gives the tonnage sent on the river to the "Saxon and Bohemian Elbe." Statistics for Internal Navigation show the quantity that crossed the Bohemian border upstream on the Elbe, practically all from Hamburg. When these figures are subtracted from the quantity sent to the Saxon and Bohemian Elbe, it is apparent that the remainder represents Hamburg's shipments to the Saxon Elbe.

Interesting results can be obtained by a proper use of these statistics. It is possible to discover how large a part the Elbe and how large a part the railway plays in

Elbe barges discharging at a steamship pier at Kuhwärder.

COMMERCE WITH THE HINTERLAND

the collection and distribution of goods for Hamburg. To be sure, we know that not all freight brought to Hamburg is destined to leave by sea; 900,000 people consume a considerable part of it and the industries which employ those 900,000 consume still more. Yet one may assume that about equal proportions of the water-brought and rail-brought freight are destined for local consumption and that the comparative significance of waterway and railway for Hamburg's seaward trade is measured by the quantity and value of the freight which the two carriers collect and distribute.

As bearing on the question of railway versus waterway transportation we are interested to find out what proportion the waterways carry of Hamburg's commerce with those sections of its hinterland which the waterways can serve: East and central Germany. Finally, we can discover the success of German railway tariffs in gaining for Hamburg shipments from West and South Germany which would normally go via the Rhine to Rotterdam and Antwerp or by rail direct to those or other foreign ports.

HAMBURG'S COMMERCE WITH THE INTERIOR, 1907.[a]

	Receipts in Hamburg		Shipments from Hamburg	
	Tons	Marks	Tons	Marks
By Elbe,	3,186,808	595,530,391	5,844,143	1,101,580,590
By Rail,	3,458,893	1,666,358,220	1,966,831	1,278,820,750
Total,	6,645,701	2,261,888,610	7,810,974	2,380,401,340

RECEIPTS AND SHIPMENTS.

	Tons	Marks
By Elbe,	9,030,050	1,697,110,980
By rail,	5,425,724	2,945,178,970
Total,	14,456,674	4,642,289,950

[a] Hamburgs Handel und Schiffahrt.

THE PORT OF HAMBURG

The customary phenomenon is observed: the waterway carries the greater part of bulky goods of low specific value, while the less bulky, more valuable articles fall to the railway. In 1907 the river trade of nine million tons, worth 1,700 million marks, represented 65 per cent of the volume, 37 per cent of the value of Hamburg's total trade with the interior. The Elbe brought 48 per cent of the tonnage, 26 per cent of the value, of Hamburg's receipts; it carried 75 per cent of the tonnage, 46 per cent of the value, of Hamburg's inland shipments. Yet this does not measure the comparative value of the Elbe as an inland carrier. That will appear later when we consider the respective parts which rail and river play in the trade with the hinterland which they serve in common: East and central Germany. The comparatively close correspondence between the totals of inland trade and the totals exported and imported in Hamburg, has already been noted.

Analyzing in more detail the figures of inland trade, we find it reflecting the nature of Hamburg's seaward trade: a preponderance among the inland shipments of raw materials for inland industries and foodstuffs for the population, a preponderance of manufactures among the receipts from inland.

HAMBURG'S INLAND TRADE BY CARRIERS AND MAIN DIVISIONS OF GOODS, 1907.[a]

RECEIPTS.

	By Elbe		By Rail		By Elbe and Rail	
	1,000 Tons	Million Marks	1,000 Tons	Million Marks	1,000 Tons	Million Marks
Foodstuffs,	1,258.9	279.3	466.2	203.2	1,725.1	482.5
Raw materials,	1,726.5	165.6	2,225.6	307.6	3,951.1	473.3
Finished products,	201.4	150.6	768.1	1,155.5	969.5	1,306.1

[a] Hamburgs Handel und Schiffahrt.

COMMERCE WITH THE HINTERLAND

SHIPMENTS.

	By Elbe		By Rail		By Elbe and Rail	
	1,000 Tons	Million Marks	1,000 Tons	Million Marks	1,000 Tons	Million Marks
Foodstuffs,	1,609.4	318.0	663.9	294.0	2,273.4	612.1
Raw materials,	4,158.9	726.0	1,105.1	717.3	5,264.0	1,443.4
Finished products,	75.8	57.5	197.7	267.4	273.5	324.9

It is observed that the Elbe predominates greatly in the carriage of foodstuffs, both according to bulk and to value. The Elbe brings the greater volume of raw materials, the railway the greater value. But nearly four fifths of the volume and seven eighths of the value of manufactured products are transported on the railway. The latter proportion is particularly large because industrial West Germany, the source of a great many exported manufactures, is not accessible to the waterway.

The part taken by Elbe and railway respectively in the distribution of the seven leading items of Hamburg's imports (see page 184), as well as the tonnage reexported, is apparent from the following table:[a]

DISTRIBUTION OF HAMBURG'S SEVEN LEADING IMPORTS, 1907.

	By Elbe		By Rail		By Sea	
	Tons	Million Marks	Tons	Million Marks	Tons	Million Marks
Coffee,	47,450	37.0	58,750	45.8	94,150	81.4
Grain,	1,107,150	141.4	47,100	5.8	499,735[b]	72.7
Wool,	25,800	44.1	53,000	95.9	32,700	49.3
Cotton,	77,450	69.7	53,900	48.5	37,300	36.5
Cow hides,	16,800	27.5	16,500	25.0	32,000	40.7
Saltpetre,	192,100	42.6	89,150	19.7	118,200	27.9
Rubber,	1,700	11.1	9,600	62.6	5,750	36.4

[a] Hamburgs Handel und Schiffahrt.
[b] Of course this is not all foreign grain.

THE PORT OF HAMBURG

In the case of four of these imports, grain, saltpetre, cow hides and cotton, the river shipped more inland than the railway; in the case of the others, wool, coffee and rubber, the railway carried more, though in the distribution of each import, excepting rubber, the Elbe carried a very respectable quantity into the interior.

A proper conception of the ability of the Elbe as an inland carrier is gained only by a comparison of its trade with East and central Germany and the rail trade with the same territory. It is fortunately possible to carry out this comparison—to compare water and rail traffic by regions. As most freight by water from Hamburg is destined for inland shipment from some upper river point, the water shipments to any section of the Elbe or its tributaries, as detailed in Hamburg's Shipping and Trade, may be considered as competing with rail shipments to adjacent traffic districts, as detailed in the Railway Statistics.

Water shipments to the German Elbe and Saale territory compete with rail shipments to traffic districts Nos. 18-20: Prussian Saxony, the kingdom of Saxony and Thuringia. Water freight to the Austrian Elbe competes with rail freight to the various traffic districts into which Austria is divided (52-55). Barges carry into the Oder territory freight that competes with direct rail shipments to Pomerania, Posen, Silesia and the other East German provinces, as well as Poland and Russia (1-4, 12-15, 50-51). Finally, water shipments to the Havel and Spree territory are in competition with rail freight direct from Hamburg to Berlin and the Mark Brandenburg (16-17). In ascertaining the proportion which waterways and railways respectively carry of the inland trade with this

COMMERCE WITH THE HINTERLAND

territory, we shall discover the true relation between railway and waterway transportation in Germany.

HAMBURG'S SHIPMENTS INLAND TO CENTRAL AND EAST GERMANY, 1907.[a]

Destination		Tons	Per Cent
I. German Elbe and Saale Territory (districts 18-20)	By Rail	405,023	15.6
	By Water	2,182,524	84.4
II. Austrian Elbe Territory (districts 52-55)	By Rail	85,753	12.
	By Water	633,800	88.
III. Oder Territory (districts 1-4, 12-15, 50-51)	By Rail	200,388	31.2
	By Water	410,727	68.8
IV. Havel and Spree Territory (districts 16-17)	By Rail	249,327	11.1
	By Water	1,991,635	88.9
Summary: Shipments to East and Central Germany	By Rail	940,491	13.5
	By Water	5,945,454[b]	86.1

Here, then, is striking evidence of the ability of waterways, properly improved and used, to compete against railways; 86.1 per cent of all freight sent from Hamburg into the competitive territory was carried by the waterways. There is no waterway connection with Schleswig-Holstein and Mecklenburg nor with West and South Germany. Two million one hundred and forty thousand eight hundred and forty-one tons of freight were sent inland to these regions by rail, 23.8 per cent of Hamburg's total shipments into the interior.

We find a similar preponderance of the Elbe in collecting goods from this competitive territory.

[a] Statistik der Güterbewegung auf deutschen Eisenbahnen, Die Binnenschiffahrt, and Hamburgs Handel und Schiffahrt.
[b] The total volume of river traffic here and on the following page differs from the figures on page 195. There the net tonnage was given, here the gross tonnage.

THE PORT OF HAMBURG

HAMBURG'S RECEIPTS FROM CENTRAL AND EAST GERMANY, 1907.[a]

		Tons	Per Cent
I. German Elbe and Saale Territory (districts 18–20)	By Rail	321,627	12.4
	By Water	2,269,380	87.6
II. Austrian Elbe Territory (districts 52–55)	By Rail	118,031	24.4
	By Water	365,000	75.6
III. Oder Territory (districts 1–4, 12–15, 50–51)	By Rail	111,433	38.3
	By Water	179,682	61.7
IV. Havel and Spree Territory (districts 16–17)	By Rail	137,508	36.
	By Water	244,495	64.
Summary: Receipts from East and Central Germany	By Rail	688,599	19.7
	By Water	3,263,443	80.3

Again we find the Elbe carrying four fifths of all goods brought to Hamburg from the competitive territory, in spite of the operation of export tariffs which are expected to keep goods on the rails. Three and six tenths million tons, or 47.8 per cent of all Hamburg's receipts from inland, came from Mecklenburg, Schleswig-Holstein, South and West Germany, with which Hamburg has only rail connection. These figures (3.6 million tons) are so large because they include one and three fourths million tons of coal and coke from the Ruhr district in Rhineland-Westphalia. This coal is for local consumption—only 125,000 tons were exported. Trade with East and central Germany constitutes the main pillar on which Hamburg's seaward trade rests. The significance of the Elbe for Hamburg is that it bears four fifths of this central and East German trade, both to and from the interior.

The Hamburg statistics designate the principal river ports with which the seaport trades, as follows:

[a] Hamburgs Handel und Schiffahrt.

COMMERCE WITH THE HINTERLAND

CHIEF RIVER PORTS TRADING WITH HAMBURG, 1907.[a]

Shipments to Hamburg			Shipments from Hamburg		
	Vessels	Tons of Freight		Vessels	Tons of Freight
Magdeburg,	1,562	522,997	Berlin,	5,325	1,470,504
Schönebeck,	789	348,098	Magdeburg,	1,649	522,761
Schönpriesen,	496	236,432	Aussig,	1,115	502,255
Tangermünde,	590	192,620	Dresden,	981	331,349
Laube-Tetschen,	580	185,080	Dessau,	833	329,660
Aussig,	324	150,497	Laube-Tetschen,	839	293,919
Dresden,	324	123,463	Schönpriesen,	630	270,911
Berlin,	1,094	101,552			

Of these river ports, Magdeburg, Tangermünde and Schönebeck lie together in Prussian Saxony on the left bank of the Elbe, Dessau on the Elbe in Anhalt, Dresden on the Elbe in Saxony. Aussig, Schönpriesen and Laube-Tetschen are on the Bohemian Elbe, just above the German border. Berlin is on the Spree.

Hamburg and Bremen are the only great German ports. They lie seventy-five miles apart on the coast of the North Sea and compete for the foreign trade of the common German hinterland. Their communication with the far eastern provinces of East and West Prussia is mostly by sea; the waterways and railways have little to do with these regions. In the trade with the rest of Germany east of the Elbe, Hamburg predominates because of the possession of the Elbe and the rivers with which it is connected. Hamburg exercises in Austria a similar preponderance, for the same reason. We have seen (page 157) that more is exported from Austria on the Elbe than via its chief seaport, Trieste. Bremen plays in all this territory an important part in the shipment inland only of goods that are a specialty of its seaward trade, particularly cotton and tobacco.

[a] Hamburgs Handel und Schiffahrt.

THE PORT OF HAMBURG

West of these regions is a territory that we may call northwestern Germany: Hannover, Thuringia and Westphalia—without the Ruhr district.[a] Bremen's inland stream, the Weser, is too short and too small (350-ton barges) to be of much service to the seaport. Bremen and Hamburg are each dependent on rail connections with this territory and both about equally distant from most of it, Bremen possessing a slight advantage in nearness to Westphalia. Yet the advantage is rendered illusory in many cases. The largest ships discharge at Bremerhafen, at the mouth of the Weser, and their freight must be sent inland by rail from there or lightered to Bremen. In either case the inland charges for freight begin at Bremerhafen, which is no nearer Westphalia than Hamburg is. In northwestern Germany the issue between the two ports is decided according to the relative strength of their seaward trade in particular goods. An analysis of Hamburg's and Bremen's inland shipments thither shows that Hamburg preponderates in wool, coffee, wine, firing wood, petroleum, tar and hides; Bremen in cotton, tobacco, rice and building wood.[b] These are goods in which the two ports have the stronger foreign trade, respectively. In all eastern, central and northwestern Germany Hamburg and Bremen have no competition from foreign seaports and none from German ports excepting in the trade with European countries: "short trading."

[a] The small but important industrial region on the Ruhr, through which the border line between the provinces of Rhineland and Westphalia runs, is divided into two traffic districts: the Rhenish and the Westphalian.

[b] See Wiedenfeld, 337-339. The figures he quotes from the German Railway Statistics for 1901 are paralleled by those for 1907.

COMMERCE WITH THE HINTERLAND

In West and South Germany we find Hamburg and Bremen subjected to severe competition. West Germany is part of the North European industrial district, a land of coal and minerals, on which have arisen the industries and the working population whose foreign trade busies the seaports of Amsterdam, Rotterdam, Antwerp, Dunkirk and Havre. As we have seen, Antwerp and Rotterdam are not only 125 to 150 miles nearer West Germany than Hamburg and Bremen are; the foreign seaports also enjoy the services of the Rhine in their communication with the West German hinterland. The Rhine is the most efficient of all waterways, both as regards the excellence of its floating stock and in the technical equipment of its river ports. In South Germany the German ports are exposed to the competition not only of Dutch, Belgian and French, but also of Austrian and Italian ports. South Germany carries on a large portion of its export and import trade via Mannheim and Ludwigshafen, twin cities on the upper Rhine that were until recently the head of navigation and were to the river what Duluth is to the Great Lakes.

There is a region where all the great European ports, both of the North Sea and the Mediterranean, can be observed in competition: lower Alsace with its highly developed textile industries and correspondingly thick population, and western Switzerland. In this region grain and petroleum are imported via Mannheim and Ludwigshafen; the cheapness of water transportation thwarts the effectiveness of even the most extreme railway tariffs for direct shipment from seaports. In cotton and coffee, the German ports, in spite of their great distance, hold their own, fortified as they are by a strong seaward

THE PORT OF HAMBURG

trade in these articles and by exceptional tariffs on German railways, designed to promote that trade. Yet even in cotton and coffee the competition of the nearer foreign ports is felt; it is overwhelming in such imports as wool, supported by no highly developed trade in German seaports and by no exceptional tariffs from them. The following table of Alsace's receipts[a] by rail in 1907 illustrates this:

ALSACE'S RECEIPTS BY RAIL, 1907.

From	Cotton Tons	Wool Tons	Coffee, Tea, Cacao Tons	Wheat Tons	Petroleum Tons
Hamburg,	1,064	4	347		116
Bremen,	21,737	254	83		197
Mannheim-Ludwigshafen,[b]	222	680	675	23,319	6,632
Holland,[c]		4	14	84	10
Belgium,	2,440	9,572	1,397	51	1,191
France,	20,556	10,630	263	193	65
Italy,	5,097	102	107		23

In the same year Switzerland received 11,400 tons of cotton from Bremen, 445 tons from Hamburg, 243 tons from Mannheim. Three thousand one hundred and three tons of coffee came from Hamburg, 185 from Bremen, 956 from Mannheim. It received 84,000 tons of wheat from Mannheim and none from German seaports; 32,400

[a] It is not possible to determine the direction of Alsace's oversea exports. Textiles sent to France, Italy or Belgium are not necessarily destined for export oversea.

[b] Mannheim-Ludwigshafen is a traffic district.

[c] These are all oversea goods, hence "from Holland" means from Rotterdam and Amsterdam; Belgium means from Antwerp; France means Dunkirk, Havre and Marseilles; Italy means Genoa.

COMMERCE WITH THE HINTERLAND

tons of petroleum from Mannheim and 650 tons from Bremen. It is impossible to discover from the statistics the part played by French and Italian ports in Switzerland. Practically, it is the success of German railway tariffs for raw cotton, supported by the oversea cotton trade in Bremen, which—in view of the importance of cotton for these textile districts—prevents them from being accounted the exclusive hinterland of French, Belgian and Dutch ports, either directly or via the Rhine. As one goes further north in West Germany, the influence of French ports disappears.

Directly in the Rhine valley, below Mannheim, the Rhine seaports, Rotterdam, Amsterdam and Antwerp, are supreme; Hamburg and Bremen are represented only by articles in which their trade is especially strong. Exports are mostly via Antwerp, with its wealth of steamship lines; imports of bulk goods—grain and ores—come primarily via Rotterdam. Amsterdam, because of its unfavorable situation, is only lightly represented.

Any considerable distance from the Rhine weakens its attractive power; it becomes cheaper to ship directly to a seaport than to tranship in a river port. West Rhineland and Lorraine, border provinces of Germany, belong wholly to Antwerp, by direct rail shipment. East of the Rhine, the river's tributary territory reaches far because of the great distance to Bremen and Hamburg: the territory of these ports begins at about Dortmund. The farther south we go, the farther east does the territory of the Rhine extend; most of Bavaria communicates with oversea via Mannheim, Frankfort on the Main, Gustavsburg and other Rhine ports. In Bavaria, furthermore, the competition of Trieste is felt and at least express

THE PORT OF HAMBURG

shipments are sent to Genoa to catch the steamers for Asia and Australia. Here, also, Hamburg and Bremen are represented primarily by the leading articles of their oversea trade, such as cotton and coffee, and they receive manufactured exports.

Considering first the Rhine valley, we find these conditions exemplified in the following statistics:[a]

BADEN'S RECEIPTS BY RAIL, 1907.

	Cotton Tons	Wool Tons	Coffee, Tea, Cacao Tons	Wheat Tons	Petroleum Tons
Hamburg,	667	78	425		265
Bremen,	17,901	18	50		96
Mannheim-Ludwigshafen,	2,307	279	857	15,582	19,278
Holland,	6	8	20		708
Belgium,	187	222	92		32
France,	1,481	45	49	10	
Italy,	924	38			

RECEIPTS OF THE BAVARIAN PALATINATE BY RAIL, 1907.

	Cotton Tons	Wool Tons	Coffee, Tea, Cacao Tons	Wheat Tons	Petroleum Tons
Hamburg,	44	38	32		36
Bremen,	3,458	112			
Mannheim-Ludwigshafen,	955	1,626	1,556	118,902	9,898
Holland,	326		30		10
Belgium,	277	2,459			
France,	3	19		10	
Italy,	130				

[a] Statistik der Güterbewegung.

COMMERCE WITH THE HINTERLAND

Receipts of Rhineland East[a] of the Rhine (without Ruhr District) by Rail, 1907.

	Cotton Tons	Wool Tons	Coffee, Tea, Cacao Tons	Wheat Tons	Petroleum Tons
Hamburg,	1,501	48	513		790
Bremen,	3,222	29	18		3,647
River ports,[b]	1,187	1,042	166	14,843	8,158
Holland,	302	37	14		55
Belgium,	560	1,367	123	10	729
France,	92	60			

Receipts by Rail of Rhineland West of the Rhine (without Saar District), 1907.

	Cotton Tons	Wool Tons	Coffee, Tea, Cacao Tons	Wheat Tons	Petroleum Tons
Hamburg,	2,862	74	5,461		2,370
Bremen,	39,680	142	292		931
River ports,	20,340	5,450	3,380	111,484	34,892
Holland,	6,471	455	8,991	1,949	1,671
Belgium,	11,344	5,386	1,032	1,310	7,635
France,	690	320	6	6	

The German ports show up more prominently as destinations of shipments from West Germany:

[a] Rhineland East of the Rhine, without the Ruhr district, is one traffic district, Rhineland West of the Rhine is another.

[b] This appears in the statistics as "local trade" (Lokalvervehr), trade with Rhineland West of the Rhine, and with the traffic district Duisburg-Ruhrort. In so far as this trade is in articles of oversea origin, like those in the tables above, it is to be ascribed to trade with the Rhine river ports.

THE PORT OF HAMBURG

Exports of West Germany, 1907.

To	RHINELAND WEST OF RHINE			RHENISH RUHR DISTRICT		
	Iron, Steel Tons	Wire Tons	Iron or Steel Wares Tons	Iron, Steel Tons	Wire Tons	Iron or Steel Wares Tons
Hamburg	10,232	353	1,442	27,037	1,051	12,704
Bremen	2,736	447	283	25,033	2,332	2,133
River ports[a]	91,600	9,567	30,775	51,386	2,294	4,002
Holland	19,137	3,364	2,894	35,152	3,123	14,572
Belgium	39,979	25,594	6,253	32,490	9,512	13,189

However, it must be taken into consideration that, of the exports to Holland and Belgium direct or with transshipment in a Rhine river port, a smaller proportion is destined for shipment oversea than of the quantities sent by rail to Hamburg and Bremen. Of the German iron exports on the Rhine, about 50 per cent are not destined for seaports.

Lorraine and the Saar district[b] show the trend of oversea foreign trade in extreme West Germany. In 1907 Lorraine received from Belgium 4,900 tons of wheat and 5,200 tons of petroleum, from Mannheim 2,185 and 1,451 tons respectively; Lorraine sent 256,900 tons of iron and steel to Belgium, 12,200 tons to Hamburg, 3,600 tons to Bremen. The Saar district shipped 90,300 tons of iron and steel to Belgium, 59,300 tons to Mannheim, 1,900 tons to Holland, 6,000 tons to Hamburg, 3,900 tons to Bremen.

In Bavaria we find the Austrian port Trieste represented:

[a] In the case of Rhineland West of the Rhine, these figures represent "local trade," trade with Rhineland East of the Rhine and trade with Duisburg-Ruhrort. In the case of the Rhenish Ruhr district the figures mean simply the trade with Duisburg-Ruhrort.

[b] The Saar region, in West Rhineland, is a separate traffic district.

COMMERCE WITH THE HINTERLAND

BAVARIA'S RECEIPTS BY RAIL, 1907.

	Cotton Tons	Wool Tons	Coffee, Tea, Cacao Tons	Petroleum Tons	Wheat Tons
Hamburg,	8,797	968	5,768	3,755	
Bremen,	55,688	134	348	1,056	
Mannheim-Ludwigshafen,	628	3,795	575	27,674	4,473
Austria,[a]	4,718	340	97	4,974	b
Italy,	11	17		9	

No better example could be found than this of the effectiveness of a strong oversea trade, backed up by favorable railway tariffs, in gaining for German seaports imports that would normally come via foreign ports and up the Rhine. Cotton imports, in the case of which these factors are present, come almost exclusively via German seaports. These factors are absent in the case of wool; it is imported in the "natural" way, on the Rhine.

The Westphalian Ruhr district imported via the river ports Duisburg-Ruhrort 49,750 tons of wheat, 14,100 tons of petroleum, 1,461,700 tons of iron ore; from Hamburg 5,800 tons of petroleum, from Bremen 328 tons of petroleum. The rest of Westphalia received 32,600 tons of wheat from the river ports, 900 tons from Bremen; but 64,000 tons of cotton from Bremen as against 560 tons from the Rhine. Similar figures are, for petroleum: Hamburg 6,800 tons, Bremen 5,300 tons, river ports 6,200 tons; for coffee: Hamburg 1,900 tons, Bremen 500 tons, river ports 55 tons. There were exported 6,000 tons of iron and steel wares to Hamburg, 9,000 tons to Bremen and 2,100 tons to the river ports. It is observed that we have again reached the tributary territory of the German seaports.

[a] Excepting Bohemia and Galicia.
[b] Wheat from Austria is not necessarily oversea wheat.

THE PORT OF HAMBURG

We have seen, then, that Hamburg's hinterland is the entire German Empire, as well as a large part of Austro-Hungary. In the special articles of its seaward trade, like coffee, oils, hides, skins, etc., the Elbe seaport reaches even into Switzerland and Russia. For connection with West and South Germany and the adjacent countries excepting Austro-Hungary, Hamburg is dependent on the railways. They have granted exceptional tariffs for nearly all German manufactures and for many of the articles of Hamburg's import trade, especially "colonial wares." This policy has been the means of extending Hamburg's hinterland far beyond the natural limits: beyond the limits of the territory for which it is the nearest great seaport.

Yet the most important part of Hamburg's hinterland is East and central Germany and Bohemia. From this territory came one half Hamburg's receipts from inland; to it were sent three fourths of its shipments. It is interesting to note that the Elbe carries four fifths of the volume of Hamburg's commerce with these, the only regions which the waterway reaches, both of the freight received and of that sent. Here is an instructive record of the power of a modern waterway to compete against a railway.

The preponderance of the river in the volume of inland freight movement is balanced by a preponderance in value of the goods that go by rail. Neither of the inland carriers has a small part to perform. The statistical tables of this inland traffic furnish proof of the efficacy of the form of water transportation and the manipulation of railway tariffs which we have described in detail in Chapter **VI**.

BIBLIOGRAPHY.

Aftalion: Le développement des principaux ports maritimes de l'allemagne. In: Revue d'économie politique. 1901.

Ausweis über den auf der österreichischen Elbstrecke von Melnik bis zur böhmisch-sächsischen Landesgrenze und deren Hafen-, Landungs- und Umschlagsplätzen im Jahre 1907 stattgefundenen Schiffahrts- und Flossverkehr.

Baasch: Geschichte der Handelsbeziehungen zwischen Hamburg und Amerika. In: Denkschrift zur Erinnerung an die Entdeckung Amerikas. Hamburg, 1892.

Barge Canal Terminal Commission of the State of New York. Report. 2 Vol., Albany, 1911. (Especially the Report from the office of Robert Skinner, U. S. Consul-General at Hamburg, Vol. I., pages 395-417.)

Boston Society of Architects. Report of Committee on Municipal Improvements. Hamburg. Boston, 1907.

Buchheister: Die Elbe und der Hafen von Hamburg. In: Mittheilungen der geographischen Gesellschaft zu Hamburg. Hamburg, 1899.

Cords: Die Bedeutung der Binnenschiffahrt für die deutsche Seeschiffahrt. Stuttgart und Berlin, 1906.

Dorn, Alexander, Ritter von Marwalt: Die Seehäfen des Weltverkehrs. Wien, 1890.

Die Binnenschiffahrt im Jahre 1907. Statistik des deutschen Reichs.

Die Entwicklung der deutschen Seeinteressen im letzten Jahrzehnt, zusammengestellt im Reichsmarineamt. Berlin, 1905.

Die Seeinteressen des deutschen Reichs, zusammengestellt auf Veranlassung des Reichsmarineamts. Berlin, 1898.

Ehrenberg: Hamburgs Handel und Schiffahrt vor 200 Jahren. In: Hamburg vor 200 Jahren. Hamburg, 1892.

Foerster: Technik der Weltschiffahrt. Berlin, 1909.

THE PORT OF HAMBURG

Goode, J. Paul: The Port of Hamburg. In: Report of the Chicago Harbor Commission. Chicago, 1908.

Grotewold: Die Ausnahmegütertarife für den Seeverkehr bei den preussischen Eisenbahnen. Bremen, 1910.

Hamburgs Handel und Schiffahrt. 1907.

Hamburgs Handel im Jahre 1907. Sachverständigen-Berichte. Hamburg, 1908.

Himer: Die Hamburg-Amerika Linie im sechsten Jahrzehnt ihrer Entwicklung. Hamburg, 1907.

Huldermann: Die Subventionen der auswärtigen Handelsflotten. Berlin, 1909.

Haarmann: Die ökonomische Bedeutung der Technik in der Seeschiffahrt. Leipzig, 1908.

Jahresbericht der Handelskammer zu Hamburg über das Jahr 1907.

Landerer: Geschichte der Hamburg-Amerika Linie. Zum 50-jährigen Jubiläum der Bestehung der Gesellschaft. Berlin, 1897.

Lehmann-Felkowski: Deutschlands Häfen und Wasserstrassen in Wort und Bild. Berlin, 1905.

Lloyd-Zeitung: Die Fortschritte des deutschen Schiffbaus. Berlin, 1909.

Löser: Die freien und Hansastädte Hamburg und Lübeck. München, 1848-49.

Mahan, F. A.: The Port of Hamburg and the Lower Elbe. Washington, 1909. (U. S. Nat'l Waterways Commission, Document No. 5.)

Nauticus: Jahrbuch für Deutschlands Seeinteressen. 1907-10.

Neumann: Die deutsche Schiffbauindustrie. Leipzig, 1910.

Passow: Das Rabattsystem der Verbände in der Seeschiffahrt. In: Zeitschrift für Socialwissenschaft, Jan., 1911.

Peters: Die Schiffahrtsabgaben. Leipzig, 1906-08.

Plath: Ansichten der freien und Hansestadt Hamburg. München, 1828.

BIBLIOGRAPHY

De Rousiers: Hambourg et l'allemagne contemporaine. Paris, 1902.
De Rousiers et Charles: Le port de Hambourg. In: Revue des questions scientifiques, 1908.
Schäfer: Article "Hanse" in: Handwörterbuch der Staatswissenschaften.
Schmidt und Kofahl: Hafen von Hamburg in Bild; eine Rundfahrt durch die Hafenanlagen... mit 170 Heliogravüren. Hamburg, 1908.
Schwarz and Von Halle: Die Schiffbauindustrie in Deutschland und im Auslande. Berlin, 1902.
Stahlberg: Der Hamburger Hafen. Seine Gliederung und sein Betrieb. Berlin, 1907.
Stahlberg: Der Hamburger Hafen und das Modell des Hamburger Hafenbetriebes in dem Museum für Meereskunde. Berlin, 1907.
Statistik der Güterbewegung auf deutschen Eisenbahnen, 1907. Berlin, 1908.
Statistische Jahrbücher für das deutsche Reich.
Thackera: Railway Freight Rates, Inland Waterways and Canals in Germany. Washington, 1911. (Document No. 19 of U. S. Nat'l Waterways Commission.)
The Picture of Hamburg, or the Englishman's Guide to that Free, Imperial City. Hamburg, 1804.
Thiess: Die Hamburg-Amerika Linie. Berlin, 1906.
Thiess: Article "Verbandsbildungen in der Seeschiffahrt," in Wörterbuch der Volkswirtschaft, 2. Auflage.
Ulrich: Preussische Verkehrspolitik und Staatsfinanzen. Berlin, 1909.
Verein Hamburger Rheeder. Bericht des Verwaltungsrats über das Jahr 1907-08.
Vereinigte Elbschiffahrts-Gesellschaften Aktiengesellschaft, Dresden. (Historisch-biographische Blätter: Industrie, Handel und Gewerbe.) Berlin, 1910.

THE PORT OF HAMBURG

Wiedenfeld: Hamburg als Welthafen. Dresden, 1906. (In Neue Zeit- und Streitfragen, 3. Jahrgang.)

Wiedenfeld: Die nordwesteuropäischen Welthäfen. Berlin, 1903.

In the footnotes, these books are referred to in abbreviated form, for instance, "Aftalion, page 24."

INDEX

Adler Line, Hapag's rate war with.............................87
Agriculture, German, insufficient for home needs................23
Alsace, commerce of, with seaports............................204
Antwerp, German exports via.............................132, 166
Antwerp, nature of its port.................................10, 34
Argentine, Hapag service to....................................94
Austria, exports of, via the Elbe..............................157
Baden, commerce of, with seaports............................206
Ballin, Albert, becomes manager of Hapag......................88
Ballin, Albert, tour of inspection to China.....................92
Barges, Elbe, in Hamburg......................................62
Barges on Elbe...146
Barge terminals..4
Barge terminals in Hamburg...................................53
Basins, in Hamburg harbor....................................46
Bavaria, commerce of, with seaports..........................209
Bavarian Palatinate, commerce of, with seaports...............206
Beer barges on Elbe..149
Beet sugar, cultivation of, in Germany.........................22
Berlin, a dependency of Hamburg.............................142
Booth Line, driven from Hamburg..............................81
Brazil, South, German settlers in..............................96
Brazilian coasting line...95
Bremen, competition of, with Hamburg........................201
Bremen, nature of its port.....................................34
British Royal Commission on Subsidies, Report of.........133, 168
Brunshausen, lightering at, formerly...........................36
Bulk goods, appearance of, in commerce.......................32
Canals, ancient, in Hamburg...................................44
Canals, disadvantages of......................................13
Channel, necessity of dredging.........................2, 31, 32
Channel, of upper Elbe.......................................145
Coal, English, in Hamburg.....................................58
Coal, in Hamburg's foreign trade..............................177
Cologne, a Hanseatic seaport..................................30
Colonies, European, rise of free trade with.....................19
Combined water-and-rail rates................................159
Consignment ports...126

INDEX

Cranes, pier, in Hamburg..........52
Cunard Line, proportion of, in American immigration..........84
Cunard Line, rate wars of, with German companies..........104
Cunard Line, does not join International Mercantile Marine Company..........108
Customs Union, German, formation of..........19
Customs Union, Hamburg enters..........21, 46
Cuxhaven, founding of..........69
Cuxhaven, preëmpted by Hamburg-American Line..........70
Cuxhaven, present significance of..........71
Cuxhaven, made a port of call..........36
Discharge, rate of, at Hamburg piers..........55
Diversity of Hapag's steamship services..........103
Docking facilities, need of..........110
Docking facilities, in Hamburg..........111
Drayage, in Hamburg..........60
Dues, channel, in Hamburg..........39
Dues, harbor and channel..........5
Dues, in various European ports..........74
Egyptian Express, of Hapag..........95
Elbe, Hamburg's inland commerce via..........196
Elbe, improvement of, lower..........37
Elbe, transportation on..........13
Elbe, upper, improvement of..........144
Elbe, value of, as inland carrier..........199
Emigrant village of Hamburg-American Line..........67
Emigrants, importance of, as return freight..........85
Emigration, German, decreases..........22
Emigration, German, to United States..........83
Emigration, South European and Russian, to United States..........84
Emigration to United States, proportion handled by German lines..........84
Exceptional tariffs on German railroads..........163
Export, leading articles of, in Hamburg..........185
Export, by countries..........187
Express steamer, type of..........118
Far East, German exports to..........132
Foreign trade and seaports..........9
Foreign trade, German, expansion of..........20, 21, 22, 80
Foreign trade, German..........27
Foreign trade, Hamburg's part in..........80
Foreign trade, German, largely with oversea countries..........24

INDEX

Foreign trade of Hamburg..................................183
Forwarders, their function in Germany.........................63
Free Port, industrial advantages of...........................49
Free Port, maritime advantages of.............................50
Free Port, origin of..48
Free Port, commercial advantages of...........................48
Free Port Warehousing Company................................64
German-East African Line founded.............................91
German-East African Line, subsidy of........................133
German Levant Line, prorating agreement of railroads with.....106
Government, German, aids foreign trade......................126
Great Lakes, freight-handling machinery on....................12
Hamburg-American Line, history of........................82 seq.
Hamburg-American Line, patron of German shipbuilding.......115
Hamburg-Australian Steamship Company, founded..............91
Hamburg, dependence on hinterland............................28
Hamburg's inland commerce, statistics of....................193
Hamburg's inland commerce, by carriers......................195
Hamburg's inland commerce, with competitive territory, by
  carriers 198
Hamburg, present channel.....................................38
Hamburg, to the sea..42
Hamburg South American Steamship Company, founded..........87
Hamburg South American Steamship Company, agreement with
  the Hapag..92
Hamburg, why it is our model..............................9, 14
Hanseatic seaports, their nature.............................15
Hapag, origin of name..82
Harbor of Hamburg, original..................................43
Harbor of Hamburg, water area of.............................59
Holland-American Line, purchased by German lines and Inter-
  national Mercantile Marine Company........................107
Holland and Belgium, canals in...............................12
Holland, her colonial power..................................18
Importer, the Hamburg.......................................125
Imports, leading, in Hamburg................................184
Imports, by countries.......................................187
Imports, distribution inland, by carriers...................197
Income from harbor dues......................................71
Income from channel dues.....................................39
Independent boatmen on Elbe.................................150
Industrial districts of Germany, distance from Hamburg.......27

217

INDEX

Industrial state, Germany becomes..............................24
Kaiser Wilhelm Ship Canal, aids Hamburg......................142
Kansas City, Mexico and Orient Railway, treaty of, with Hapag..93
Kingsin Line, founded..86
Kingsin Line, bought by the Hapag..............................88
Kosmos Line, founded...87
Kosmos Line, agreement with the Hapag.........................89
L. C. L. shipments, assembled by forwarders...................164
Levant, Hamburg's exports to..................................167
Levant, line to, from New York.................................90
Lightering 61
Lightering, formerly...44
Lighters, sea-going, plying from Hamburg.......................96
Liners, types of, in Hamburg...................................35
Liverpool, nature of its port............................10, 12, 34
Lloyd, North German, patron of German shipbuilding..........114
Lloyd, North German, Chinese coasting and river lines of........92
Lloyd, North German, subsidy of...............................130
London, financial center of the world.........................125
London, nature of its port............................10, 12, 34, 61
London, the old order of seaport.................1, 3, 12, 15, 18, 20
Lübeck, head of the Hansa......................................15
Lübeck, remains of its prestige................................17
Mediæval seaports, location of.................................30
Merchant carrier, the..80
Merchant Marine, Hamburg......................................98
Merchant marine, Hamburg, growth of.........................179
Merchants in Hamburg, their activity.........................121
Merchants in Hamburg, their position in the oversea world......126
Merchants, the seaport..123
Merchants, function of..8
Midland Canal..141
Mississippi, similarity of, to Elbe.............................12
Mooring posts...46, 56
Museum für Meereskunde, Hamburg pier in model of............54
New York, nature of port of...................................34
New York, similarity of, to Hamburg...........................11
Oil barges on Elbe..149
Pier sheds..52
Piers, leased..57
Piers of the Hamburg-American Line............................51
Piers, state, lines at..57

218

INDEX

Pools and agreements, between German companies............105
Pools and agreements, between German companies and International Mercantile Marine Company......................106
Population, German, how its increase is cared for............25
Prorating with steamship lines, by German railroads..........137
Protective tariff, enacted in 1879............................21
Quay walls, first, in Hamburg................................45
Railroad connection, in river ports..........................154
Railroads, coöperation of, with the Elbe.....................156
Railroads, Hamburg's inland commerce by......................196
Railroad rates, aid of, to German seaports...................162
Railroad rates, inland and export, compared..................165
Railroad terminals in Hamburg.................................62
Rates, railroad..5
Re-exportation in Hamburg....................................188
Register ton, definition of...................................31
Rhine, German exports via...............................166, 169
Rhineland, commerce of, with seaports........................207
Rhine river ports, service to, from Hamburg...................97
River ports, Elbe, trade of, with Hamburg....................201
River ports, terminal facilities in..........................154
Sailing vessels replaced by steamers..........................86
Sailing vessels, size of, in the forties.....................32
Saxon railroads, coöperation of, with the Elbe...............157
Shipbuilding, in Germany.....................................111
Shipbuilding, in North Sea yards.............................112
Shipbuilding, effect on, of North German Lloyd subsidy.......113
Shipowners, function of, in a seaport.........................8
Shipping, in German ports, statistics of.....................173
Shipping, in North Sea and Baltic ports......................174
Shipping, in great European seaports.........................176
Shipping, in Hamburg....................................177, 181
Ship types, in the merchant marine...........................117
Ship types, of the Hapag and the Lloyd.......................119
Shipyards, chief German......................................116
Silesia, a dependency of Hamburg.............................142
South America, trade with....................................181
South Germany, trade of, with Hamburg...................170, 203
Steamers, freight, on Elbe...................................148
Steamship lines from various European ports..................100
Steamship lines, principal, in Hamburg...................98, 180
Steamship lines, in the world................................100

INDEX

Steamship lines, their importance to a port.....................4, 77
Steamship lines, significance of....................................4
Stettin Lloyd, driven from Hamburg............................81
Stettin, high channel dues of....................................41
Suez Canal, draught of..38
Subsidies, various sorts of, in Germany.....................127, 129
Subsidies, of various merchant marines........................136
Terminal facilities, railroad connection...........................2
Terminal facilities in river ports..............................154
Terminal facilities for barges in Hamburg......................152
Third-class passage on Hamburg-American Line.................119
Tolls on the Elbe, formerly.....................................39
Towboats on Elbe..147
Tramp steamers..78
Transit tariffs on railroads....................................171
Transshipment, Hamburg's dependence on, formerly..............78
Union Steamship Company, Hapag's rate war with...............88
United Elbe Navigation Company..........................149, 151
Warehouses in Hamburg...64
Water rates, on Elbe..158
Waterways and railways, coöperation between.....................7
Waterways, their significance to a seaport......................140
Waterway system of Hamburg..................................140
Waterway transportation, theory of..............................6
West Africa, line to, from New York............................90
West Germany, trade of.......................................170
West Germany, trade of, with Hamburg....................203, 208
West Indies, competition in, between Hapag and Royal Mail
    Steamship Company...93
Westphalia, Peace of, decentralizing effect of....................19
Woermann Line, founded.......................................89

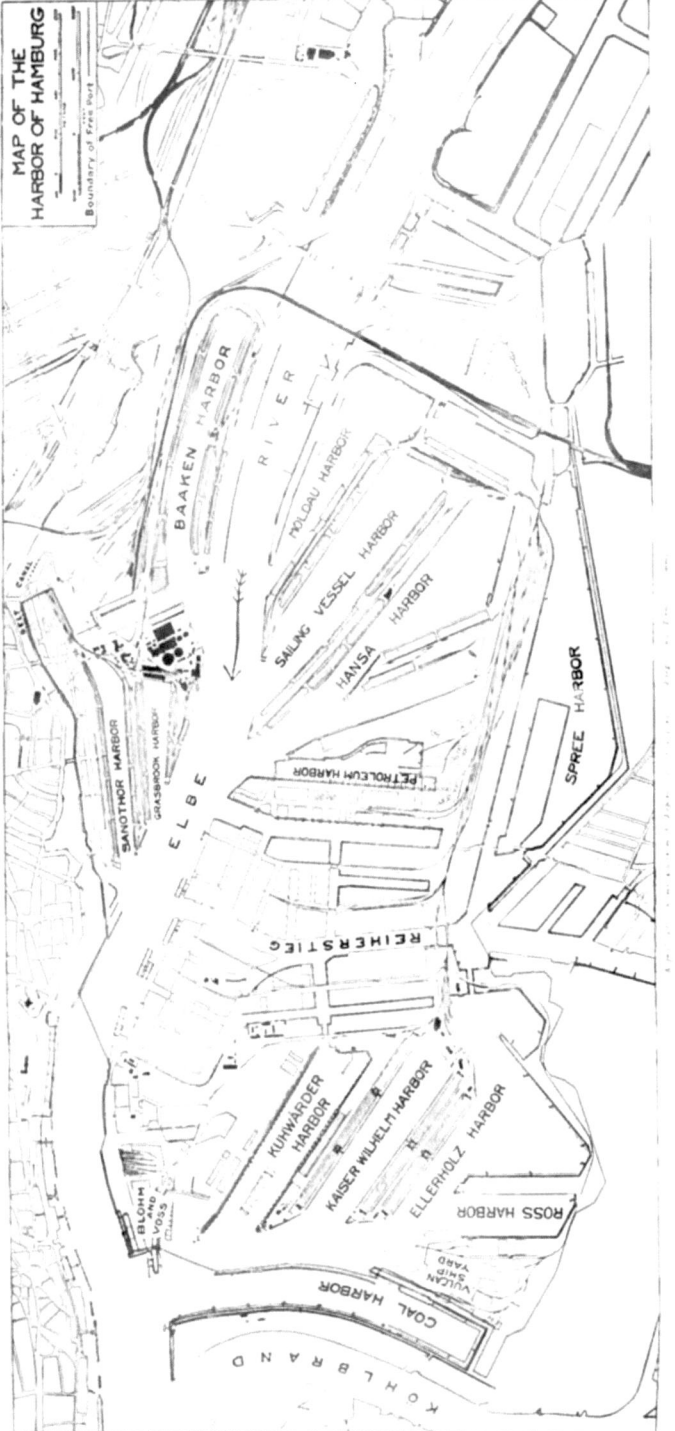